Digital Data Integrity

Digital Data Integrity

The Evolution from Passive Protection to Active Management

DAVID B. LITTLE
SKIP FARMER
OUSSAMA EL- HILALI
Symantec Corporation, USA

John Wiley & Sons, Ltd

Published in 2007 by John Wiley & Sons Ltd, The Atrium, Southern Gate, Chichester,
West Sussex PO19 8SQ, England

Email (for orders and customer service enquiries): cs-books@wiley.co.uk
Visit our Home Page on www.wiley.com

Other Wiley Editorial Offices

John Wiley & Sons Inc., 111 River Street, Hoboken, NJ 07030, USA

Jossey-Bass, 989 Market Street, San Francisco, CA 94103-1741, USA

Wiley-VCH Verlag GmbH, Boschstr. 12, D-69469 Weinheim, Germany

John Wiley & Sons Australia Ltd, 42 McDougall Street, Milton, Queensland 4064, Australia

John Wiley & Sons (Asia) Pte Ltd, 2 Clementi Loop #02-01, Jin Xing Distripark, Singapore
129809

John Wiley & Sons Canada Ltd, 6045 Freemont Blvd, Mississauga, ONT, L5R 4J3, Canada

Anniversary Logo Design: Richard J. Pacifico

British Library Cataloguing in Publication Data

A catalogue record for this book is available from the British Library

ISBN 978-0-470-01827-9 (HB)

Typeset in 10/12 pt Sabon by Thomson Digital
Printed and bound in Great Britain by Antony Rowe Ltd, Chippenham, Wiltshire
This book is printed on acid-free paper responsibly manufactured from sustainable forestry
in which at least two trees are planted for each one used for paper production.

Contents

Acknowledgements

I would like to dedicate this effort to my wife Cheryl, my son Tarik, and my daughter Alia. I am also especially grateful to my parents Mohammed Larbi (1909–1996) and Zakia Sultan (1927–2006).

– Oussama El-Hilali

A big thanks to my father, Charles, for his support and advice. Our discussions helped me to remain focused, I guess this is a long way from our homework discussions in my younger days. My mother, Serene, and girlfriend, Laurette Dominguez, always had encouraging words and offered support at all the right times. And thanks to my grandmother, Fannie Bigio, who always said 'nothing ventured, nothing gained', for reminding me that anything is possible.

– Skip Farmer

I want to first thank my wife, Nancy, for all her support during this long and sometime arduous process. We can not accomplish much without a supportive family behind us and I am no exception. My kids, Dan, Lisa, Jill, Jeff and Amanda, have always been there as well as my parents, Ray David and Jeffie Louise Little. Thanks to you all. I am sure that my family and my co-workers were beginning to wonder if there really was a book. I guess this is the proof that once again, there is light at the end of the tunnel. This book would never even have happened without the support of Brad Hargett and Bryce Schroder who afforded me the time as needed. The original driver again behind this entire project was Paul Massiglia. I would also like to thank Richard Davies, Rowan January and Birgit Gruber from Wiley UK who have shown us tremendous patience and have offered us a lot of help. Last, but certainly not the least is my thanks to God; it is only by the strength of Christ that I am able to do anything. Thank you.

– David B. Little

We would like to thank all those who helped with this book especially Paul Mayer, Ray Shafer and Wim De Wispelaere for their valuable contributions. We would also like to thank Rick Huebsch for allowing us to use NetBackup documentation.

Dave Little, Skip Farmer and Oussama El-Hilali

Introduction

We would like to welcome you to share our views on the world of data integrity. Data protection has been an unappreciated topic and a pretty unglamorous field in which to work. There were not a lot of tools to assist you in setting up a data protection system or actually accomplishing the task of providing true data protection. The attitudes have been changing lately due to a number of technology trends such as the low cost of disks, increasing availability of high bandwidth and computation power. As a result, analysts such as Gartner are predicting a change in the role of the IT organization and its potential shift from a cost center to a value center. We are going to look at this subject from the viewpoint of overall data protection and how we are seeing data protection and data management merging into a single discipline. We will start with a brief walk down memory lane looking at the topic of data protection as it has existed in the past. We will also take a look at some of the data management tools that are being commonly used. We will then look at how these two formerly separate tool sets have started coming together through necessity. We will also highlight some of the factors that are driving these changes. We will then take a look at what we think the future might hold.

We have attempted to keep this book as vendor neutral as possible and provide a generic look at the world of data protection and management. The one area where we have used a specific product to explain a technology is in Chapter 4 where we talk about bare metal restore (BMR). In this chapter, we have used Symantec Corporation Veritas NetBackup Bare Metal Restore™ to demonstrate the BMR functionality.

1   OVERVIEW

In this book, we will chronicle the traditional backup and recovery methods and techniques. We will also go through some of the other traditional data protection schemes, discussing how the paradigm has shifted from the simple backup and recovery view to the one of data protection. From here we will go into some of the changes that have been occurring and give some of the reasons that these have been happening. There is discussion on some of the traditional data management methodology and how people have tried to use this to either replace or augment their data protection schemes. New data protection applications have already started to integrate some of these processes and these will be discussed along with the new data protection features that are emerging in the marketplace. We will also take a look at some of the methods used to protect the actual integrity of the data. This will include encryption and methods to control access to the data.

2   HOW THIS BOOK IS ORGANIZED

This book is presented in two parts. The first part, Data Protection Today, consists of Chapters 1–6. In these chapters, we will take a look at the way data protection has been traditionally accomplished. Chapter 1 looks at traditional backup and recovery along with hierarchical storage management and how it can augment the overall data protection scheme. We also take a look at disaster recovery and management challenges. Chapter 2 looks at some of the traditional disk and data management tools. This includes the different RAID (redundant array of independent (inexpensive) disks) technologies as well as replication. In Chapter 3, we get the first glimpse of the future, the integration of the protection and management methodologies. We will examine the ways the disk tools are being leveraged by the backup applications to provide better solutions for you the consumer. Chapter 4 takes a close look at the problem, and some of the solutions, of BMR. We close part 1 with a look at management, reporting, and security and access in Chapters 5 and 6.

In part 2, Total Data Management, we look at where things are going today and our view of where they are going tomorrow, at least in the realm of data integrity. Chapter 7 gives us our first look at some of the exciting new features that are being offered for data protection.

Chapter 8 examines the rapidly growing arena of disk-based protection technologies. Chapters 9 and 10 look at the changing requirements around management and reporting and the tools that are evolving to meet these requirements. We close this part with a look at some of the tools that are becoming available for the total system, including the next generation of BMR, true provisioning and high availability.

Of course, we will also offer a table of contents at the beginning and an index at the end, preceded by a glossary and an appendix or two. We hope that these tools will allow you to determine what areas of the book are of most interest and can help guide you to the appropriate sections. We tried not to write a great novel, but rather provide some information that will be helpful.

3   WHO SHOULD READ THIS BOOK

In this book, we address a large audience that extends from the general reader to the practitioner who is involved in implementing and maintaining enterprise wide data protection and data management systems and processes. By discussing today's state of data protection, we expose some of the technologies that are widely used by large enterprises and comment on user issues while offering our views and some practical solutions. At the same time, we talk about new issues facing the future enterprise as a result of shifts in business practices or discovery and adoption of new technologies.

Whether it is tools or techniques, the general reader will find in this book a good set of discussions on a vast array of tools such as hierarchical storage manager (HSM), BMR and techniques like mirroring, snapshots and replication. The reader will also find in this book a good summary of some of the advanced technologies like synthetics, disk staging and continuous data protection.

The practitioner will find in this book an exploration of user and vendor implemented solutions to cope with today's complex and ever demanding data protection needs. The designer and architects who are deploying new systems or redeploying existing data protection infrastructures will enjoy our reflections on what works today and what does not. They can also benefit from the technical description of new technologies such as single instance store (SIS) that are surfacing today in data protection and setting the stage for this industry to be a part of data management in the future.

4   SUMMARY

By combining technical knowledge with day-to-day data protection and data management issues, we hope to offer the reader an informative book, a book that is based on knowledge as well as observation and reflection that emanates from years of experience in developing data protection software and helping users deploy it and manage it.

Chapter 1

An Introduction to Data Protection Today

1.1 INTRODUCTION

As we start our discussion of the future of data protection, we would like to spend some time taking a look at data protection today and establishing some of the basic terminology that is commonly used. This will be a review for most, but it helps avoid confusion with some of the terms and usage. It also helps set the groundwork for looking to the future. We will start out this discussion by looking at the traditional backup and recovery.

1.2 TRADITIONAL BACKUP AND RECOVERY

When we talk about data protection today, we usually talk about the traditional backup and recovery, generally, the process of making secondary copies of production data onto tape medium. This discussion might also include some kind of vaulting process. This has been the standard for many years and to an extent continues to meet the foundational requirement of many organizations; that being an ability to recover data to a known-good point in time following a data outage, which may be caused by disaster, corruption, errant deletion or hardware failure. There are several books available that cover this form of data protection, including *UNIX Backup and Recovery* by W. Curtis

Digital Data Integrity   David Little, Skip Farmer and Oussama El-Hilali

Preston (author), Gigi Estabrook (editor), published by O'Reilly and *Implementing Backup and Recovery: The Readiness Guide for the Enterprise* by David Little and David Chapa, published by John Wiley & Sons. To quote from the very first chapter in *Implementing Backup and Recovery: The Readiness Guide for the Enterprise*, 'A *backup* is a copy of a defined set of data, ideally as it exists at a point in time. It is central to any data protection architecture. In a well-run information services operation, backups are stored at a physical distance from operational data, usually on tape or other removable media, so that they can survive events that destroy or corrupt operational databases.'

The primary goals of the backup are to be able to do the following:

- Enable normal services to resume as quickly as is physically possible after any system component failure or application error.
- Enable data to be delivered to where it is needed, when it is needed.
- Meet the regulatory and business data retention requirements.
- Meet recovery goals, and in the event of a disaster, return the business to the required operational level.

To achieve these goals, the backup and recovery solution must be able to do the following:

- Make copies of all the data, regardless of the type or structure or platform upon which it is stored, or application from which it is born.
- Manage the media that contain these copies, and in the case of tape, track the media regardless of the number or location.
- Provide the ability to make additional copies of the data.
- Scale as the enterprise scales, so that the technology can remain cost effective.

At first glance this seems like a simple task. You just take a look at the data, determine what is critical, and decide on a schedule to back it up that will have minimal impact on production, install the backup application and start protecting the data. No problem, right? Well, the problem is in the details. Even the most obvious step, determining what is the most critical data can be a significant task. If you ask just about any application owner about the criticality of their data, they will usually say 'Mine is the most important to the organization.' What generally must happen is that you will be presented with various analysis summaries of the business units or own the task of interviewing the business unit managers yourself in order to have them determine the data, the window in which backup may run, and the retention level of the data once it is backed up. What you are doing is preparing a *business impact analysis* (BIA). We will discuss the BIA later in this chapter when we discuss disaster recovery (DR) planning. This planning should yield some results that are useful for the policy-making process. The results of these reports should also help define the recovery window, should a particular business unit suffer a disaster. The knowledge of these requirements may, in fact, change the budget structure for your backup environment, so it is imperative during the design and architecture phase that you have some understanding of what the business goals are with regard to recovery. This can help you avoid a common issue faced by the information technology (IT) staff when architecting a backup solution, paying too much attention to the backup portion of the solution and not giving enough thought to the recovery requirements. This issue can easily result in the data being protected but not available in a timely manner. This issue can be compounded by not having a clear understanding of the actual business requirements of the different kinds of data within an enterprise which will usually dictate the recovery requirements and therefore the best method for backing up the data. You should always remember that the primary reason to make a backup copy of any data is to be able to restore that data should the original copy be lost or damaged.

In many cases, this type of data protection is actually an afterthought, not a truly thought-out and architected solution. All too often when a data loss occurs, it is discovered that the backup architecture is flawed in that the data was either not being backed up at all or not being backed up often enough resulting in the recovery requirements not being met. This is what led us to start recommending that all backup solutions be architected based on the recovery requirements. As mentioned above, BIA will help you avoid this trap.

When you actually start architecting a backup and recovery solution as a part of the overall data protection scheme, you start looking at things such as

- Why is the data being backed up?
  - Business requirements.
  - Disaster recovery (DR).
  - Protection from application failures.

- Protection from user errors.
- Specific service level agreements (SLAs).
- Legal requirements.
- What is the best backup strategy to meet the recovery requirements?
  - Backup frequency.
  - Backup type: full, differential incremental or cumulative incremental.
  - Data retention.
  - Off-site storage of images.

As you look at all these different elements that are used to make the architectural decisions, you should never loose sight of the fact that there is usually an application associated with the data being backed up and the total application must be protected and be recoverable. Never fear, the true measure of a backup and recovery system is the restorability of the data, applications and systems. If your backup and recovery solution allows the business units to meet or exceed their recovery SLAs, you will get the kind of attention we all desire.

Although a properly architected backup and recovery solution is still an important part of any data protection scheme, it is becoming apparent that the data requirements within the enterprise today require some changes to address these new requirements and challenges. Some of the changes are

- total amount of data;
- criticality of data;
- complexity of data, from databases, multi-tier applications as well as massive proliferation of unstructured data and rich media content;
- complexity of storage infrastructure, including storage area networks (SAN), network attached storage (NAS) and direct attached storage (DAS), with a lack of standards to enforce consistency in the management of the storage devices;
- heterogeneous server platforms, including the increased presence of Linux in the production server mix;
- recovery time objectives (RTO);
- recovery point objectives (RPO).

These requirements are starting to stress the traditional data protection methodology. The backup and recovery applications have been adding features to give the data owners more tools to help them address these issues. We will discuss some of these in the following chapters.

1.3   HIERARCHICAL STORAGE MIGRATION (HSM)

HSM is another method of data management/data protection that has been available for customers to use and is a separate function from tradition backup, but it does augment backup. With a properly implemented HSM product that works with the backup solution, you can greatly reduce the amount of data that must be managed and protected by the backup application. This is accomplished by the HSM product managing the file system and by migrating off at least one copy of inactive data to secondary storage. This makes more disk space available to the file system and also reduces the amount of data that will be backed up by the backup application. It is very important if implementing an HSM solution to ensure that the backup product and the HSM product work together so that the backup product will not cause migrated files to be recalled.

A properly implemented HSM application in conjunction with a backup application will reduce the amount of time required to do full backups and also have a similar effect on the full restore of a system. If the backup application knows that the data has been migrated and therefore only backs up the placeholder, then on a full restore only the placeholders need to be restored. The active files, normally the ones you are most concerned with, will be fully restored and restored faster as the restore does not have to worry with the migrated inactive data. Retrieving migrated data objects from nearline or offline storage when an application does access them can be more time consuming than accessing directly from online storage. HSM is thus essentially a trade-off between the benefits of migrating inactive data objects from online storage and the potentially longer response time to retrieve the objects when they are accessed. HSM software packages implement elaborate user-definable policies to give storage administrators control over which data objects may be migrated and the conditions under which they are moved.

There are several benefits of using an HSM solution. As previously stated, every system has some amount of inactive data. If you can determine what the realistic online requirements are for this data, then you can develop an HSM strategy to migrate the appropriate data to nearline or offline storage. This results in the following benefits:

- reduced requirements for online storage;
- reduced file system management;
- reduced costs of backup media;
- reduced management costs.

HSM solutions have not been widely accepted or implemented. This is mostly due to the complexity of the solutions. Most of these applications actually integrate with the operating system and actively manage the file systems. This increases the complexity of implementing the solution. It also tends to make people more nervous about implementing an HSM product. This is probably one of the least understood product of the traditional data protection and management products.

1.4   DISASTER RECOVERY

Another key ingredient of the traditional data protection scheme is DR. In the past, this was mostly dependent on a collection of backup tapes that were stored either at a remote location or with a vaulting vendor. In many instances, there was no formal planning or testing of the DR plan and procedures. As you might expect, many of these plans did not work as desired. Recently, more emphasis has been given to DR and more people are not only making formal plans but also conducting regular DR tests to ensure that they can accomplish the required service levels. We have always said that until your DR plan is tested and demonstrated to do what is needed, you do not have a plan at all.

As stated earlier in this chapter, do not succumb to the temptation to concentrate too much on the raw data and forget about the overall production environment that uses the data. If the critical data exists within a database environment, the data itself will not do you much good without the database also being recovered. The database is of only marginal value if all the input comes from another front-end application. As you put together a DR plan, you should always try to remember the big picture. Too often people concentrate on just recovering specific pieces without considering all the interdependences. By developing the BIA mentioned earlier you can avoid a lot of the potential pitfalls. One of the interesting results of gathering the proper data necessary to do the BIA can be a change in the overall way you architect backup and recovery for your enterprise. An example of this is a customer who discovered they were retaining too much data for too long a period of time due to lack of a business analysis of the data looking at both it's immediate value, the effects time had on the value of the data, and the potential liability of keeping too much data around too long. After doing the BIA the customer reworked their retention policy and actually experienced a sizeable cost savings by putting cartridges back into circulation.

The BIA is basically a methodology that helps to identify the impact of losing access to a particular system or application to your organization. This actually is a process that is primarily information-gathering. In the end, you will take away several key components for each of the business units you have worked with, some of which we have listed here:

1. Determine the criticality a particular system or application has to the organization.
2. Learn how quickly the system or application must be recovered in order to minimize the company's risk of exposure.
3. Determine how current the data must be at the time of recovery.

This information is essential to your DR and backup plans, as it describes the business requirements for backup and recovery. If you base your architecture on this information and use it as the basis for your DR plan, your probability of success is much greater. Another by-product of the BIA and the DR plan is developing a much better working relationship between the business units, application owners and the IT staff.

With the growing emphasis on DR and high availability, we begin seeing the mingling of data protection and data management techniques. Users started clustering local applications and replicating data both locally and remotely. We will discuss these in detail in a later chapter. RTO and RPO requirements are two key elements to consider when making the decision on which technique to use for DR, as seen in Figure 1.1.

As history has shown us, there are many different kinds of disasters, and a proper DR plan should address them. The requirements can be very different for the different scenarios. There is an excellent book that can be very helpful in preparing a good DR plan. It is *The Resilient Enterprise: Recovering Enterprise Information Services from Disasters* from VERITAS Software publishing.

1.5 VAULTING

Any discussions that concern DR should also include a discussion about the vaulting process. In very basic terms, a vaulting process is

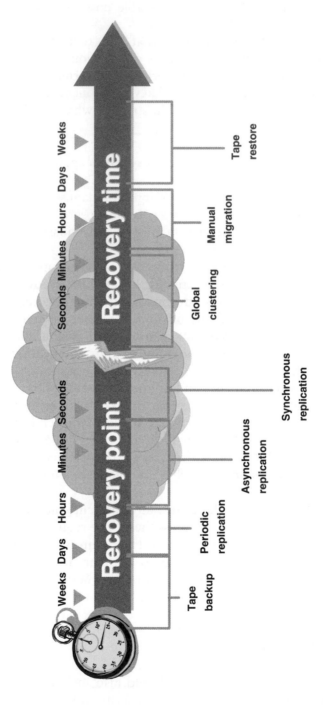

Figure 1.1 RPO/RTO

the process that allows you to manage and accomplish any or all of the following steps:

- Create a duplicate of the backup image for storage off-site.
- Automate ejecting the media that need to be taken off-site.
- Provide reports that allow you to track the location of all backup media.
- Manage recalling media from the off-site location, either for active restores or for recycling after all the data on the media has expired.

It is possible to develop all the tools and procedures to accomplish all of these tasks, but it can be a tedious and potentially risky endeavour. Some of the backup applications offer a vaulting option that is fully integrated with the backup solution, which is much easier to use and more reliable. Figure 1.2 shows the basic vaulting process flow.

There are at least three options for creating a backup image that will be taken off-site for secure storage:

- Take the media containing the original backup images off-site.
- Create multiple copies of the backup image during the initial backup.
- Have the vaulting process duplicate the appropriate backup images.

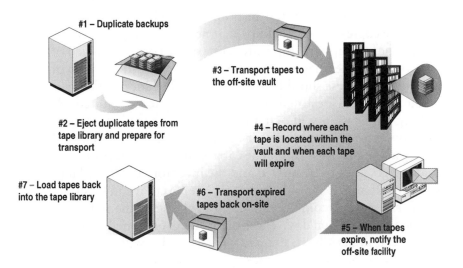

Figure 1.2 Basic vaulting flow

1.5.1 Offsiting Original Backup

If you select this method of selecting which medium is stord offsite in a secure storage facility you must be prepared to accept a potential delay in restore requests. Any request to restore data will require the original media be recalled from the storage facility. This obviously not only will affect the time to restore but also has the real possibility of causing the backup media to be handled more, which can reduce the life of the media. It also puts you at a greater risk of loosing media as it is being transferred more often.

1.5.2 Create Multiple Copies of the Backup

Some of the backup applications have the ability to create more than one copy of the backup data during the initial backup process. By doing this, you can have your vault process move one of the copies off-site. This removes the problem of always having to recall the off-site media to fulfil all restore requests. It also makes the off-site copy available as soon as the backup is completed.

1.5.3 Duplicate the Original Backup

This has been the more common method of creating the off-site copy of the backup. After the initial backup is complete, the vault process will create copies of any backups that need to have an off-site copy.

After the backups are duplicated, one of the copies is moved off-site. After you have the images on media that are ready to be taken off-site, the vaulting process should create a list that includes all the media IDs for all the media destined to be taken off-site or vaulted. A good vaulting application will actually perform the ejection of the media, so the operator or administrator can physically remove the media.

The vaulting process should be capable of creating reports that show what images need to be moved and the inventory of all media that are currently off-site. It should also create a report that can be shared with the off-site storage company that shows all the media that need to be returned on any given day. These are generally the media on which all the backup images have expired. These media will be recalled and reintroduced into the local backup environment, usually going back into an available media set.

A good vaulting application will also manage the backup and off-site storage of the data that makes up the backup application's internal catalogue. It will also track this information, which is very important if you need to recover the backup server itself.

The off-site storage companies have warehouses that are especially built for providing the highest possible protection against disasters – natural and otherwise. These companies offer services to physically transport the tapes to and from the warehouse. Some advanced vaulting applications provide reports and data formats that make it easy to integrate with the vault vendor's own data management systems. It is important to remember that backup is a sticky application. Users should carefully evaluate the potential off-site storage vendor for their staying power. Backup is also a critical application, so the user should look at what support the vendor is able to provide. You want to be comfortable that the backup vendor and the off-site storage company are going to be around for the long haul. Otherwise, all those backup images that the user has been saving for 7 years might be of little use.

1.6   ENCRYPTION

There is a rapidly growing requirement that all data that is moved off-site be encrypted. The data protection application vendors are hurriedly working on updating the existing encryption solutions to allow for more selective use. The entire subject of encryption is detailed in Chapter 6, but we can highlight some of the requirements and options that are currently available:

- Client-side encryption.
- Media server encryption.
- Encryption appliance.

1.6.1   Client side encryption

With client-side encryption, all of the data that is moved from the client is encrypted before being sent off the client. This involves using the client central processing unit (CPU) to actually perform the encryption. This can have a performance impact on the client, depending on how much of the CPU is available and therefore can have an impact on the backup performance.

1.6.2   Media server encryption

This method of encryption allows you to encrypt only backups that are being sent off-site or just those being created by the vault process. This still uses a CPU to perform the encryption, but now it is the media server CPU that is being used. The basic work of the media server is that of a data mover and generally there is not as high a demand on its CPU. You also have more control on when this is being done so you can pick a more idle time. The downside here is that the data is moving across the network from the client without being encrypted.

1.6.3   Encryption appliance

This method involves purchasing a specialized hardware appliance that is installed in the data stream. This appliance can encrypt data as it passes through. It removes the CPU load of the other two methods, but does require the purchase of the special hardware with its own software/firmware.

As we will see in Chapter 6, the process of encrypting the data is only a piece of the puzzle. Generally when you elect to encrypt data there are keys involved that must be managed. Without the proper keys the data becomes very secure. No one can read it, not even the owner. The key management is different for each of the options.

1.7   MANAGEMENT AND REPORTING

In the traditional backup and recovery data protection scheme, there is generally a silo approach to management with each group doing its own management and reporting. This duty usually falls on the administrators, not the people who actually need the information. This just becomes another task for administrators who have plenty of other responsibilities. In many cases, they do not actually know the SLAs that they are reporting on.

Reports are typically generated by scraping the application logs and presenting either the raw data or some basic compilation of the data being collected. The resulting reports often do not have enough details or the correct details to facilitate the type of management that is truly required to ensure that all the SLAs are being met. The fact that we often have the wrong people trying to manage the data protection scheme with inadequate reporting has made overall data protection too often not properly implemented and managed.

This is further compounded by the fact that reports concerning storage are generally done by the storage administrators, reports concerning systems by the system administrators and reports about the network by the network administrators. It is very difficult for any one person or group to know exactly how well the enterprise is being managed and protected with this widely diverse method of management and reporting.

1.7.1 Service Level Management

Increasingly, storage services, including backup and recovery, are offered to business unit 'customers' based on established service levels. The business units are then charged back based on their consumption of the resource, bringing a measure of accountability into IT resource

Table 1.1 Service levels

	Availability shelf life	Recovery time	Recovery point	Backup window	Underlying technologies
Platinum	Forever	5 min	Zero data loss	No impact 24 × 7	Data replication, snapshots and off-host backup to tape on replica data off-site
Gold	7 years	10 min for first 30 days, 1 h for next 11 months, 1 day after that	30 min	No impact 24 × 7	Rolling snapshots every 30 min with 24 h retention, 1 year worth of nearline tape capacity in library, on shelf for remainder
Silver	2 years	1 h for the first year, 1 day for the second year	1 h	No impact during production day, midnight to 6 a.m.	
Bronze	30 days	24 h	24 h	Backup can occur anytime	Daily backup to tape

consumption. Service levels can generally be established into a small number of narrowly defined offerings, based upon the metrics by which a business unit has recoverability. The metrics are not communicated in IT terms, such as server platform, tape or disk technology, SAN/NAS and so on, but rather in simple terms that quantify the expectations for data recovery. For example, one could establish a simple four-tier hierarchy, which offers platinum, gold, silver and bronze services. An example of service levels is shown in Table 1.1.

By establishing clear SLAs and monitoring delivery against these commitments, the operation can be properly funded by more responsible business unit owners. Also, the underlying technology infrastructure can be better managed and upgraded as needed to allow the storage group to deliver on its commitments to the business units.

1.8   SUMMARY

As we have seen, historically data protection has been accomplished by traditional backup and recovery with some mingling of HMS solutions. This was coupled with DR schemes that were also mostly based on the same backup and recovery techniques and included a vaulting process. The silo approach to reporting did little to assist in moving beyond this methodology. We are starting to see service levels also becoming a part of the management process.

In the following chapters, we will see the move that has already started to augment these traditional data protection techniques with the more tradition data management tools. In later chapters, we will follow some of the more advanced integration of these tools and techniques and then look beyond these to the totally new approaches being developed to meet the data protection needs of today and tomorrow.

Chapter 2
The Evolution

2.1 INTRODUCTION

Almost all of the data in the enterprise today exists on some kind of spinning storage, typically a disk. As disk management tools advanced, people started relying on them for data protection. This involves both hardware implemented tools and server-based volume managers. We will take a look at some of these basic tools and highlight the short-comings of relying on them for data protection. This will be a review for most, but helps establish a good basis for our further discussions.

2.2 STORAGE VIRTUALIZATION

We should first have a basic discussion about storage in general. When we talk about data being on the disk and being managed by disk management tools, where exactly is the data? We can best categorize data as existing in one of the four places:

- Internal disk(s) – Disk drives that are physically inside the cabinet of the server.
- Standalone – A disk drive that is in its own enclosure.
- JBOD (just a bunch of disk) – Disks that share an enclosure but have no intelligent interface.
- Array – Two or more storage devices that are in a common enclosure with some intelligent control and are managed by a common body of the control software.

Digital Data Integrity   David Little, Skip Farmer and Oussama El-Hilali
© 2007 Symantec Corporation. All rights reserved

Data that is located on the internal disks, standalone disks or JBOD is traditionally managed by a host-based volume manager if you want to implement RAID (redundant array of independent (inexpensive) disks) or replication. For data located on an array, you can use either a host-based volume manager or the internal control software that is a part of the array. The host-based volume manager or the internal control software of the array actually provides you the capability of creating virtual storage. Once you have created the virtual storage, you can then apply the desired disk management techniques.

Virtual storage devices do not really exist. They are simply representations of the behaviour of physical devices of the same type. These representations are made to the application programs or operating system in the form of responses to I/O requests. If these responses are sufficiently like those of the actual devices, you need not be aware that the devices are not 'real'. This simple but powerful concept is what makes all of storage virtualization work – no application changes are required to reap its benefits. Any application or system software that can use disk drives can use equivalent virtual devices without being specifically adapted to do so.

2.2.1 Why Storage Virtualization?

Why would you want to do this? Data storage devices are simple devices with straightforward measures of quality. A storage device is perceived as good if it performs well, does not break (is highly available), does not cost much and is easy to manage. Virtualization can be used to improve all four of these basic storage quality metrics:

I/O performance. More and more, data access and delivery speed determine the viability of applications. Virtualization can stripe data addresses across several storage devices to increase I/O performance as observed by applications.

Availability. As society goes increasingly online, tolerance for unavailable computer systems is decreasing. Data can only be 'there' if data storage is equally 'there'. Virtualization can mirror identical data on two or more disk drives to insulate against disk and other failures and increase availability.

Cost of capacity. Disk storage prices are decreasing at an amazing rate, which in part accounts for equally amazing increases in storage consumption. Virtualization, in the form of mirroring and remote replication, increases consumption still further. Thus, the cost of delivered storage capacity remains a factor. Enterprises can exploit lower storage costs either as reduced information technology spending or as increased application capability for a constant spending level. Virtualization can aggregate the storage capacity of multiple devices or redeploy unused capacity to other servers where it is needed, in either case enabling additional storage purchases to be deferred.

Manageability. Today's conventional information technology wisdom holds that the management of system components and capabilities is the most rapidly increasing cost of processing data. Virtualization can combine smaller devices into larger ones, reducing the number of objects to be managed. By increasing failure tolerance, it reduces the downtime, and therefore, the recovery management effort.

The case for storage virtualization is simple. It improves these basic measures of storage quality, thereby increasing the value that an enterprise can derive from its storage and data assets.

Actually, even a single disk drive is virtualized today. This virtualization is accomplished by firmware within the physical disk drive. No one has to worry about sector, track and head layouts. Virtualized I/O interfaces allow disk technology to evolve without significant implications for users. A disk drive might be implemented using radically different technology from its predecessors, but if it responds to I/O commands, transfers data and reports errors in the same way, support implications are minor, making market introduction easy. The virtualized I/O interface concept is embodied in standards such as small computer system interface (SCSI), advanced technology attachment (ATA) and Fibre Channel Protocol (FCP). Disk drives that use these interfaces are more easily introduced into production environments enabling applications to immediately exploit the benefits they deliver. There is much more discussion today around the storage virtualization of disk arrays.

2.3   RAID

RAID has become very popular in the data center. RAID is used to enhance I/O performance, data availability and manageability. Essentially, all enterprise storage systems incorporate some kind

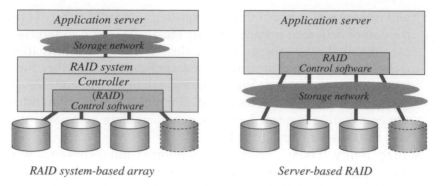

Figure 2.1   Hardware-based and server-based RAID systems

of RAID. Host-based volume managers also provide RAID and often offer the capability to further enhance I/O performance by combining the capacity of two or more storage systems. The challenge for the system architect or administrator is how to best choose from the multiple forms that are available. An example of RAID is shown in Figure 2.1.

2.3.1   So What Does This Really Mean?

The acronym RAID is defined in more detail as follows:

Redundant means that a part of the devices' storage capacity is used to store check data. Check data is information about the user data that is redundant in the sense that it can be used to recover the data if the device that contains it becomes unusable.

Array simply refers to the set of devices managed by the control software that presents their net capacity as one or more virtual storage devices. The control software that is typically called a volume manager or logical volume manager runs on the host. In disk systems (commonly called RAID systems), the control software runs in specialized processors within the systems.

Independent means that the devices are capable of functioning (and failing) separately from each other. RAID is a family of techniques for combining ordinary storage devices under common management.

Disks are the physical disk drives whose storage capacity is virtualized.

2.4   RAID LEVELS

There were five basic levels of RAID initially documented, but levels 2 and 3 are not often seen as they require special-purpose hardware.

RAID level 1 – mirroring. The mirroring technique consists of making two or more identical copies of each block of user data on separate devices. Mirroring provides high data availability at the cost of an extra storage device (and the adapter port, enclosure space, cabling, power and cooling capacity to support it). Most mirrored volumes deliver somewhat higher read performance than equivalent volumes that are not mirrored and only slightly lower write performance.

RAID level 4 – parity. This RAID level uses large stripes, which means you can read records from any single drive. This allows you to take advantage of overlapped I/O for read operations. As all write operations have to update the parity drive, no I/O overlapping is possible on writes. RAID 4 is best suited for sequential data access and is not seen very often.

RAID level 5 – parity. This RAID level interleaves check data (in the form of bit-by-bit parity) with user data throughout the array. At times, a parity RAID array's storage devices operate independently, allowing multiple small application I/O requests to execute simultaneously. At other times, they operate in concert, executing one large I/O request on behalf of one application. Parity RAID is suitable for applications whose I/O consists principally of read requests. Many transaction processing, file and database serving and data analysis applications are in this category.

Figures 2.2 and 2.3 show some of the different RAID levels just discussed.

In addition to these, the term RAID level 0 has come into common use to denote arrays in which user data block addresses are striped across

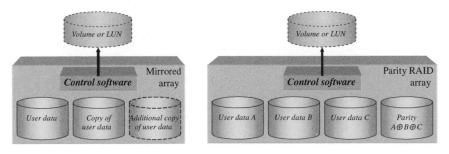

Figure 2.2   Mirrored versus parity RAID

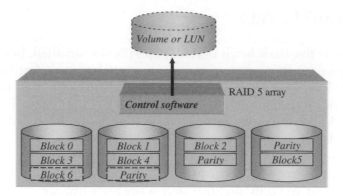

Figure 2.3   RAID level 5

several devices, but in which there is no check or parity data. Striped arrays provide excellent I/O performance in almost all circumstances, but do not protect against the loss of unreadable user data. An illustration of RAID level 0 can be seen in Figure 2.4.

There are also layered RAID levels, which are being used. These are referred to by the RAID level corresponding to the levels being combined. The most common are

RAID level 10 – a combination of RAID level 0 and RAID level 1. This can be RAID level 0 + 1, where the data is striped and then the stripes are mirrored, or RAID level 1 + 0, where the data is mirrored and the mirrors are then striped across multiple devices.

RAID level 50 – a combination of RAID level 5 and RAID level 0. This type consists of a series of RAID level 5 groups that are striped

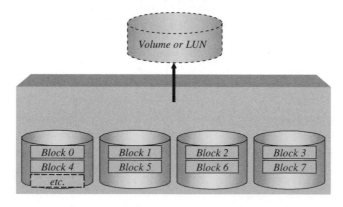

Figure 2.4   RAID level 0 – striped array

in RAID level 0 fashion to improve the RAID level 5 performance without reducing data protection.

All types of RAID except level 0 have two distinguishing features:

- *Provide failure tolerance through data redundancy.* RAID arrays hold redundant information about user data in the form of check data. Check data enhances user data availability by enabling recovery of user data blocks that have become unreadable. For example, mirrored virtual disks use one or more complete copies of user data as check data; parity RAID virtual disks use a parity function computed on several corresponding user data blocks.
- *Convert virtual device block addresses to physical storage device block addresses.* The most common forms of conversion used in RAID arrays are concatenation and striping, the latter of which enhances I/O performance by balancing most I/O loads across some or all of an array's storage devices. Whether mirrored or parity RAID, block storage virtualization systems typically stripe data across devices in a regular geometric pattern, or offset user data block addresses by some fixed amount, or a combination of the two.

RAID is very popular primarily as it offers protection against disk spindle failures (expect for RAID 0) and can also offer better I/O performance. The other advantage of RAID is that it presents multiple devices to the operating system and the application as a single logical unit. What has been the primary failure of RAID as a part of data protection is that it offers no protection for data corruption or user errors. There is an excellent book that goes into great detail about this entire subject, *Virtual Storage Redefined* by Paul Massiglia and Frank Bunn, published by VERITAS Software.

Tales From the Real World

An example of the downfall of relying on RAID for true data protection was a customer who had a very critical application that had all the data on a mirrored array. Not only was the application data mirrored, but the operating system for the server was also mirrored. One evening, the server crashed as can happen with just about any server. After trying several times to reboot, the customers decided to use the mirrored boot disk. Mirroring protects against failures, right? After spending a couple of hours trying to boot the system from the mirrored boot disk, they finally decided to revert to the procedures to recover a failed server. They eventually discovered that the original problem was a system failure that caused the operating system to be corrupted. The mirroring worked just fine. The corruption was mirrored in real time and both copies of the operating system were destroyed. They lost several hours of very valuable time.

2.5   WHAT MIRRORING AND RAID DO NOT DO

Mirroring and RAID reduce the probability of online data loss significantly, but focus almost entirely on disk drive failures. Failure of other storage system components can cause data loss as well:

- *Network cables, I/O adapters and interfaces.* These are the pieces that form a path to a storage device. Failure of one of these elements may block communication with several devices. If storage is configured so that only one device in an array is connected to any given path, path failure is survivable. If not, path failure results in unavailable data (but not necessarily lost data unless there are accompanying device failures).
- *Power and cooling systems.* A failed power supply causes all the devices it powers to be unavailable. Fan failure eventually destroys devices by overheating. Power supplies and fans are usually configured in redundant pairs, each powering or cooling several disk drives, and sized so that one can adequately power or cool the companion drives if its companion should fail.
- *External RAID controllers and appliances.* Failure of the RAID controller or storage appliance through which several storage devices are accessed makes the devices inaccessible and is generally not regarded as an optimal design. RAID systems and storage appliances should be configured in pairs that connect to the same storage devices and application servers. When everything is functioning normally, I/O load is shared. When one fails, its partner takes control of all devices and executes all I/O requests.
- *Embedded RAID controllers.* From a storage access standpoint, failure of an embedded RAID controller is equivalent to failure of the computer in which it is embedded. Some vendors have improved on this situation by devising controllers that interact with operating systems to fail over disk drives from a failed embedded controller to a second controller in a different server.
- *Application servers.* Except in the case of host-based volume managers, a server failure is not precisely a failure of the I/O system.

Increasingly, however, business requires that applications resume quickly after a server failure. This need has given rise to clusters of servers that support each other, with a designated alternate taking over a failed server's work. Server failover has slightly different impact on different types of storage systems. Volume managers and embedded RAID controllers must take control of a failed server's drives, verify array consistency and present arrays as virtual devices to the applications in the alternate server. External RAID systems and storage appliances must present the failed server's drives to the alternate server, typically on different I/O paths. A server failure does not affect the internal consistency of these arrays, although consistency of file systems or databases on them must be verified.

In addition to these storage system component-related factors, there are at least two additional important potential causes of data loss:

- human error;
- application fault.

Mirrored and RAID arrays store blocks of binary data reliably, regardless of its meaning. It must be noted that a RAID array stores incorrect data just as reliably as it stores correct data. Mirroring and RAID do not protect against data corruption due to human errors or application faults. A combination of quality data managers (e.g. logging file systems and database managers) and regular backup with an enterprise backup and recovery application offers the only realistic protection against these causes of data loss.

We have learned that protection against storage device and other component failures is necessary, but not totally sufficient for high data availability. The entire I/O system, as well as servers and applications, must be protected against both physical and logical failures. Mirroring and RAID are building blocks for highly available data access, but not for the entire solution.

2.5.1 Which RAID Should I Use When?

With all the different RAID levels available, it is sometimes confusing as to which level to use and when to use it. Table 2.1 shows some of the pros and cons to the different RAID levels, which might help with these decisions.

Table 2.1   RAID level pros and cons

RAID level	Pros	Cons
RAID 0	Excellent performance	No fault tolerance
RAID 1	Excellent read performance	Slightly lower write performance
	Most fault tolerant	Double disk requirement
	Can be used for off-host processing	
RAID 4	Good read performance for sequential access	Low write performance
	Protects against single device failure	Data loss if two disks fail
	Less disk required than RAID level 1	Cannot be used for off-host processing
RAID 5	Good read performance	Better performance than RAID level 4
	Protects against single device failure	Data loss if two disks fail
	Less disk required than RAID level 1	Cannot be used for off-host processing
RAID 10 (0 + 1)	Good performance unless a fault occurs	Requires double disk
	Fault tolerant	Performance degrades more with failure
RAID 10 (1 + 0)	Good performance	Requires double disk
	Most fault tolerant	
RAID 50 (5 + 0)	Better performance than RAID level 5	Less performance than RAID level 1
	Protects against single device failure	Data loss if two disks fail
	Less disk required than RAID level 1	Cannot be used for off-host processing

2.6   REPLICATION

Replication has become very popular as a part of the highly available (HA) architecture. We find that sometimes people confuse HA with data protection. HA is defined as the ability of a system to perform its function continuously (without interruption) for a significantly longer period of time than the reliabilities of its individual components would suggest, whereas data protection implies the ability to have the data recoverable regardless of a failure or disaster. Some people are now including replication as a part of their overall data protection scheme. Replication gives the user the ability to maintain identical physical copies of a master set of data at two or more widely separated locations. Some of the reasons to use replication are:

- *Distribution and consolidation.* Distribution of data from a central data center to remote locations is a common data processing strategy. Examples of distribution include regional web sites and catalogues maintained by global enterprises. Similarly, some enterprises require periodic consolidation of data accumulated at remote offices to a central data center. In this case, data from many sources is replicated to a single target location.
- *Off-host processing.* It is often necessary to process a snapshot of data without impacting applications that are using the live data. The most common example of this is to provide data for a backup application. Business considerations often make it impractical to stop or even impede application execution for any appreciable period. Even if snapshots are used, backup or analysis overhead can lead to unacceptable application performance. However, if data is replicated to another server, that server can back up or analyse data with no impact on live application performance.
- *Disaster recovery (DR).* Perhaps, the most important use of data replication is to maintain copies of the most business critical data at remote locations for quick recovery from disasters that incapacitate entire data centers and make local high-availability mechanisms ineffective.

When distances between data source and target are short, split mirrors can meet these requirements. Often, however, the geographic separation between source and target precludes mirroring. In these cases, data replication is often the most viable alternative. Data replication technology has evolved to meet these needs. The technology differs from mirroring in that it takes three important factors into account:

- *Latency.* The time to communicate between an application server and the location at which data is replicated may be significant. Increasing the application response time by the wait for remote writes to complete may be unacceptable.
- *Communication reliability.* Communications between an application server and a replication site may occasionally fail. Brief outages should be transparent to applications. More extensive recovery procedures should only be required for lengthy outages (tens of hours).
- *Source to target.* Unlike mirroring technology, in which all mirrors are regarded as equal, replication has a definite source and target.

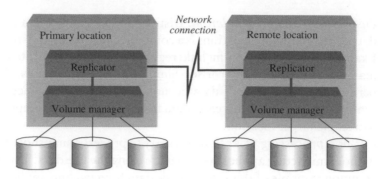

Figure 2.5   Basic host-based replication

These factors lead to data replication designs that differ considerably from mirroring, even though they deliver similar results, at least over short distances. Replication can be performed by the hardware as a part of a disk array offering or by a host-based software volume manager. In Figure 2.5, we see an example of a basic host-based replication.

When we discuss replication, we must keep in mind that there are different types of replication, each with its own advantages and disadvantages. The basic types of replication are

- *Synchronously*, with write request completion not signalled to applications until updates have been transmitted and written to persistent storage at the replication target location.
- *Asynchronously*, with write request completion signalled to applications as soon as updates have been logged on persistent storage at the source.

Synchronous replication can have a performance impact on the application, but many of the replication vendors have developed methodologies that lesson the impact. The advantage is that you do not have to be concerned about the order of the writes from the primary to the secondary nor you have to worry about losing data at the secondary site. With asynchronous replication, there is much less impact on the application, but there is more risk in that a system crash or long network interruption has the potential of causing data loss. Most replication vendors have developed tools to lesson the possibility of this happening. Before selecting what kind of replication is best suited for your needs, you will want to make sure you understand how the proposed solution will handle all the possible situations.

2.6.1   Host-Based Replication

Host-based replication software intercepts application write requests to replicated virtual devices and sends them with their data to target locations where complementary replication software writes them to persistent storage. The advantages of host-based replication are:

- *Storage independence.* With host-based replication, storage devices at the source and target locations may be different. This allows repurposed or lower cost storage devices to be employed at target locations.
- *Enterprise network sharing.* Although dedicated network connections may be configured for performance, host-based replication is TCP/IP based and can therefore share network facilities with other traffic. This can reduce both the expense and the complexity of managing replication.

Host-based replication technology generally requires that source and target platforms be of the same type. They can replicate data between storage devices of different types, but only if the devices are connected to servers of the same type.

2.6.2   RAID System Replication

Like host-based replication, RAID system replication has no information about the meaning of data in the virtual device blocks being replicated, so virtual devices are the only objects that can be replicated by RAID systems. The source RAID system itself sends write requests made to virtual devices to a companion RAID system at the target location, where they are written to virtual devices of equivalent capacity. With today's technology, source and target RAID systems must generally be identical. However, there are reports of hardware RAID systems that will allow replication between similar but not necessarily identical platforms. The data copying phase of RAID system replication is transparent to both source and target servers.

Some RAID replication systems require dedicated connections between the source and the target. These systems replicate data with no application overhead. The application server starts and stops replication and controls data recovery, but during replication, application writes at the source are copied transparently to target devices with minimal impact on the application performance. The principal advantages of RAID system-based replication are

- *Minimal application server resource.* Though dedicated communications can be costly, minimal application impact is an offsetting factor. Some RAID systems support replication over TCP/IP networks, eliminating the expense of a dedicated replication network.
- *Application server platform independence.* A single enterprise RAID system at the source can replicate data for multiple server platforms of different architectures to one or more target locations.

RAID system-based replication typically requires that both source and target RAID systems are of the same model. Thus, a RAID system can replicate data for different server types, but only between storage devices of the same type.

2.7   STANDBY OR DR SITE

With the greater acceptance of replication, the next logical progression in the data management and data protection scheme was to implement standby or DR sites. Initially, this was done by contracting with a company that specialized in providing equipment for just this purpose on a rental basis. More and more, enterprise companies and agencies have multiple data centers with each center being the standby for another one. This allows for more control of the systems and more control of the data. Replicating block storage device contents places up-to-date (or nearly up-to-date) data at a DR site, but it does not necessarily guarantee that the devices containing the data are in a fit state to be used when a disaster occurs at the main data center. The problem arises because disasters can occur at awkward times such as during a file system metadata update or a database manager cache flush, for example.

Before the replicas can be used, the file systems on them must be mounted and restored to a consistent state, databases must be recovered by database manager recovery tools and any other application-specific crash recovery actions must be taken. Thus, the process of restarting services at a recovery site after a disaster is similar to that of restarting services after a crash at the main data center, provided that updates are replicated in the same order as they are applied at the main data center.

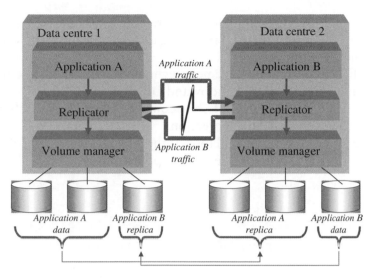

Figure 2.6   Bi-directional replication

Even if you only have two data centers you can use this approach as noted in Figure 2.6. You can actually replicate critical data between the two centers, so if either goes down the other can pick up the critical applications.

An often overlooked part of the standby or DR site strategy is what to do after you fail over to the other site. Actually, the failover part of the process is the easiest to architect and implement and test. The much larger part of the problem is figuring out how to fail back when the primary location is back in operation. Any data that has been created at the standby or DR site must now be 'moved' back to the primary location. If you have configured backups to occur at the alternate location, then this data must also be a part of the consideration for the return to normal production strategy. Also, if the primary site is permanently unavailable, you need to have a plan to replicate this process to a third standby or DR site. If you are considering this kind of strategy, we cannot mention strongly enough that you must plan and test and test and test. Until you have successfully done an exhaustive test, you cannot say you have a viable plan.

2.8   SUMMARY

In this chapter, we have looked at some of the most common disk or data management tools that are being used for data protection. We actually started with a discussion of the overall topic of disk virtualization. We then looked at RAID including a look at the different levels of RAID most commonly used.

Another of the data management techniques we took a look at was replication and the use of standby and DR sites.

We examined the potential shortcomings all of these tools have when it comes to true data protection.

In the next chapters, we will look at the integration of some of these same tools with backup and recovery applications.

Chapter 3
Backup Integration

3.1 INTRODUCTION

With the explosion in the amount of data being protected, people have started looking for alternatives to the traditional data protection techniques. Some people have been trying to leverage disk management tools for data protection to augment backup and recovery but have found these also not to be adequate, mostly due to the inherent lack of overall protection. Many backup application vendors have started delivering tools that combine the data management/disk management tools and the traditional backup techniques. In this chapter, we will look at some of the more common integrations.

3.2 SNAPSHOTS

A snapshot is a point in time copy of the data. It is also called a frozen image as it represents a nonchanging view of the data at the time the snapshot is created. There are two types of snapshots that we will discuss: mirrors and copy-on-write (COW) snapshots.

- *Mirrors* – A mirror is an identical copy of the data on two or more disks or sets of disks.
- *COW snapshot* – A COW snapshot consists of a list of blocks whose contents have changed since snapshot initiation and a private data area containing the blocks' contents prior to the change. When you read a COW snapshot, blocks unmodified since snapshot initiation

Digital Data Integrity   David Little, Skip Farmer and Oussama El-Hilali

are read from their original locations. Modified blocks' prior contents are retrieved from the snapshot's changed block area. The net effect is that the backup (and other applications that access the snapshot) sees the image of data as it stood at the instant of snapshot initiation.

3.2.1   Mirror

Mirroring is one of the oldest tools used to help protect data. The mirror can be created and managed by the disk hardware or by a host-based volume manager. Mirroring has been widely used to protect against a disk hardware failure. There is no protection against data corruption or user errors. This has limited the value of mirrors as a primary data protection method, but still mirrors are very common, especially with relational databases as they offer protection for hardware failures. As the data continues to grow in size and importance, users and data protection application vendors have begun to look at ways to leverage the use of mirrors. This has basically led to the following two approaches:

- mirror as an instant recovery mechanism;
- mirror as a backup object.

3.2.1.1   Mirror as an instant recovery mechanism

One of the methods being used is to create a mirror and then split off one copy. This copy is then reserved for reporting applications or other usages that are read only. In reality, this is a standby copy of the original data that is available in case the original data is corrupted or deleted. One of the primary considerations in this method is that the mirror is no longer an exact copy of the data once the split occurs. If you need to resort to this copy of the data, you must be able to potentially move back in time to the point where the split occurred. This is the recovery point objective (RPO). The time criticality and the volatility of the data will generally determine how often the mirrors will be synchronized. In most cases, this happens at least once a day.

3.2.1.2   Mirror as a backup object, either by the application
server or by a backup server

The other use of mirrors as a part of the overall data protection is to use a mirror as a backup object. Although this approach can be similar to the first method, often there are differences. In this case, either a custom script or the backup application will quiesce the data and break the mirror. The application will continue to write to the primary data volume, whereas the secondary copy is used as the backup object. This lessens the impact of the backup on the application and also lessens some of the pressure on the backup window. As this method has become more popular, it has been further enhanced. Now, some of the backup applications offer the capability to not only break off a copy of the data but also logically move this secondary copy of the data via the split mirror to a backup server which will then actually perform the backup. This allows the backup process to have even less impact on the application as the application server is not performing the backup. This truly helps remove the pressure on the backup window. The following figures show how this is accomplished. Figure 3.1(a) shows the basic configuration required. Figure 3.1(b) shows the result of logically moving the data to the backup host. In figure 3.1(c), you can see that this also allows a single backup client to be used to perform the backups for multiple application hosts.

3.2.1.3   Mirror resynchronization

After the split mirror has been used for either purpose, it must be resynchronized so that it can be used again. There are two basic techniques that are used:

- *Full content copy*. The complete contents of the parent virtual device are copied to the newly rejoined mirror.
- *Fast resynchronization*. If all changes to a mirrored virtual device are logged from the moment at which a mirror is split for read-only use until the moment at which it is rejoined, then the rejoined split mirror can be resynchronized by copying only changed blocks from its original mirror. This technique obviously requires that the split mirror's contents not be modified while it is in the split state. Fast resynchronization typically completes much more quickly and with much lower resource utilization than a full content copy. Although fast resynchronization of split mirror snapshots causes less resource contention than a full content copy, both do contend with applications for I/O resources during resynchronization. A good practice when using split mirrors is to choose rejoin times that will have minimal effect on the application performance.

The other important consideration with split mirror snapshots for the backup is the capital cost. Each mirror requires as much physical storage as the original volume. A terabyte mirrored volume requires two terabytes of physical storage (devices, enclosures, power and cooling, interconnects and floor space), a three-copy mirror requires three terabytes and so forth. Although the decreasing cost of storage makes this less of a barrier, it is nonetheless a factor when choosing a snapshot backup strategy. An example of fast resynchronization is shown in Figure 3.2.

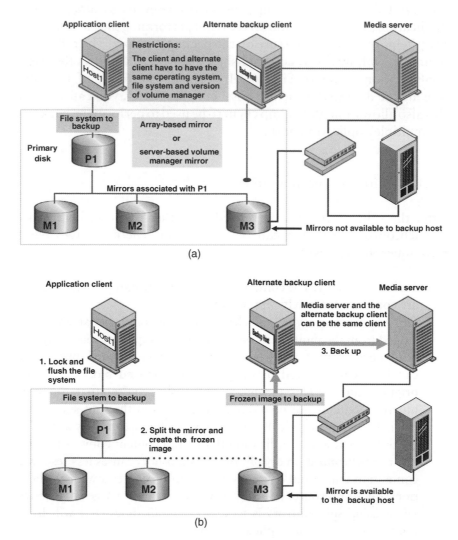

Figure 3.1   (a–c) Alternate client backup of split mirror

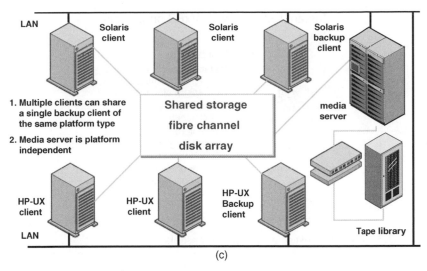

(c)

Figure 3.1   (*Continued*)

3.2.2   COW Snapshot

The COW snapshot is attractive for two reasons:

- It consumes less physical storage than mirror snapshots. In fact, the COW snapshot storage requirements are equal to the original volume size plus unique block changes that occur while the snapshot exists.

Fast mirror resynchronization

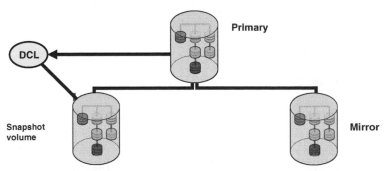

Snapshot volume = mirror detached/split from the primary
Mirror = identical copy of a primary
DCL = data change log (DCL)
 - created when the mirror was detached to become a snapshot volume
 - stores updates to the primary as the associated snapshot volume was created
 - used by the mirror FastResync to update a snapshot volume, making it identical to the primary, i.e. a mirror

Figure 3.2   Fast mirror resynchronization

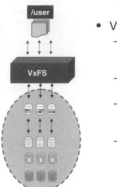

Figure 3.3   Instant recovery using VxFS snapshots

As it is typical that small fractions of large data sets change during backups, the COW technique can enable an administrator to keep several snapshots online. This process is now being called instant recovery backups as the snapshots are instantly available in the case of corruption or application failure. An example of this is seen in Figure 3.3. This particular example is shown using the Symantec Veritas file system VxFS as the snapshot tool. The snapshots are being managed by Symantec Veritas NetBackup.

- It requires little or no resynchronization overhead. When a COW snapshot has served its purpose, the storage space it occupies is simply returned to whatever free space pool it came from (a file system's free space pool or a raw volume used as a cache device). There is no resource contention comparable to that resulting from split mirror snapshot resynchronization.

COW snapshots have unique limitations as well:

- The most obvious is that backup and applications contend for the same physical I/O resources. The less the change to a virtual device for which snapshots are in effect, the more the contention will exist, because applications and backup contend for the same data images. The copied contents of modified blocks can be kept on different storage devices, but the backup and applications will access the same unmodified data blocks.
- Application updates of live data take longer when a COW snapshot is in effect. The first application update of a block after snapshot

initiation requires that the block's prior contents be read and rewritten to a changed block area. Once a block has changed and its prior image has been saved, the snapshot imposes no further incremental overhead. The more unique the application update of data blocks while COW snapshots are in effect, the more the impact will be on the overall write performance.

- COW snapshots do not protect against data loss if their parent volumes are destroyed or become inaccessible. Locating a file system and its COW snapshots on fault-tolerant devices may provide this protection. Unlike a split mirror snapshot, a COW snapshot is no longer useful if its parent is destroyed.

A COW snapshot can be used as a backup object as it does represent a frozen image of the data. A COW snapshot cannot be logically transported to an alternate backup host as it is not a complete copy of the original data.

3.2.3   Replication

Replication has become very popular as a part of the highly available (HA) architecture. Some people are now including replication as a part of their overall data protection scheme. As we saw in Chapter 2, replication is similar to mirroring without distance restrictions. Backup applications are also looking at ways to leverage remote replication of data. One of the ways this is being accomplished is to have the backup application that is running at the primary site trigger a mirror split at the remote site and then initiate a backup of the split mirror at the remote site. An example of this can be seen in Figure 3.4.

This approach does require that the servers in both locations be running the same operating system with the same volume management software. It also requires that both be in the same backup domain as the backup server at the primary site controls the backup operations on the server at the remote site. We expect to see variations on this methodology to become more prevalent in the future. This type of solution offers a lot of advantages. You have the HA aspects of replication plus the true protection against data deletion, user error and corruption that is not offered by replication alone. It can also meet the requirement for off-site storage of backup images as the backup images are actually being created at an off-site location away from the primary data location.

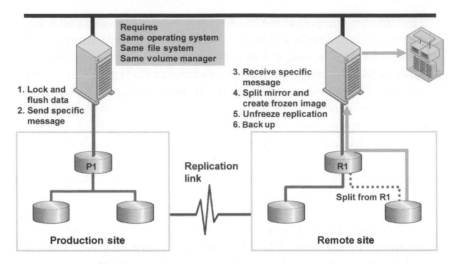

Figure 3.4   Replication with backup

Another growing use of replication and the backup and recovery application is to actually replicate the backup master server environment including all the appropriate application database information to a second site. This allows you to start recovery operations much more quickly in the event of a site-level disaster. Normally in the event of a disaster and failover to a remote site, the backup master server must be recovered from a copy of the catalogue backup which was sent along with the off-site duplicate backup volumes. For relatively large catalogues, this step alone may take hours. There may also be other preliminary processing that must occur on the backup master server. These operations must precede any application server data recovery.

Using a replication technology to replicate the master server environment can minimize or even eliminate this delay. The nearly real-time replication of the catalogue can help eliminate the time-consuming step of recovering the master server from tape. Combining replication with snapshot technology can make it possible to mount a copy of the primary catalogue at the remote site and execute any processing which must precede application restores.

3.2.4   Applications

It is generally accepted that most of the data today is kept in or managed by some kind of application, most often a relational database.

The backup application vendors are constantly developing agents that interface with the business applications, especially those that have an interface that allows it to be quiesced. This enables the backup application to more easily perform 'hot backups' which are backups of the business application data that can be performed while it is still online and available to the users and customers. Some of these business applications are very mature in this area and offer a lot of backup options.

Today, it is common to use the more advanced integrated data protection features we discussed earlier in this chapter for protecting the business application data. Some of the business application vendors are extending these capabilities by adding even more application interfaces or more options for the backup agents to leverage. Almost all of the business application data today exists on some type of RAID device. With the backup vendors offering the ability to manage, via the business application interface – the pausing of the business application, the creation of a snapshot of the data, releasing the business application and then using the snapshot as a backup object, we see the perfect marriage of different tools and techniques to allow for better data protection.

3.2.5   Summary

We have taken a look at some of the ways disk and data management tools are starting to be integrated by the backup applications. We looked at snapshots which included mirrors as well as COW snapshots. These are becoming very popular as a part of the data protection scheme as they both offer unique backup and recovery options.

We also took a further look at replication which we started discussing in Chapter 2. This is another of the tools that are being integrated to offer unique backup and recovery capabilities.

As we finish up looking at integration, we have to mention the applications themselves as that is generally where the data comes from. As we stated, the business application developers are also looking at the problem of data protection and at the typical management tools and offering more and better interfaces to protect their data.

Chapter 4

Bare Metal Restore

4.1   INTRODUCTION

As we have seen through the first three chapters, data protection tools
and techniques have been changing as everyone tries to keep up with
data growth and the ever-increasing dependence of companies and
organizations on their data. In this chapter, we will look at the emerging technology of bare metal restore (BMR). We use the term *bare
metal* to refer to the process by which the entire system is recovered
onto hardware that does not yet have an operating system installed.
Although we will use this term generically in this chapter, we should
also mention that Symantec NetBackup Bare Metal Restore™ is also a
product name and trademarked term registered to Symantec Corporation. To avoid confusion between the product Symantec NetBackup
Bare Metal Restore™ and the generic term bare metal restore, we will
use the trademark symbol (™) with the term when referring to the
Symantec product. As we explore the topic of BMR, we will use the
Symantec Corporation product NetBackup Bare Metal Restore™ to
explain this technology.

4.2   BACKGROUND

Data protection solutions are implemented primarily so that data can
be recovered when the need arises. Although data protection solutions
were designed to recover one or more, or even all of the files within a file
system, they were not primarily designed to recreate the file systems

Digital Data Integrity   David Little, Skip Farmer and Oussama El-Hilali

themselves nor were they designed to recover the operating system. These capabilities are required for the bare metal recovery of a server which has suffered a catastrophic failure.

4.2.1   Why BMR?

Why is it important to have a BMR capability integrated into the data protection solution? First, without this integration, you must quite literally piece together the system from different protection methods. Second, when you back up all of the data on a system, you want to be able to use that data to recover the system. Although you may expect that this capability should exist in every data protection solution, we have only recently begun seeing BMR capabilities integrate with enterprise data protection applications.

Among the primary considerations for the design of a data protection solution to guard against a server failure or site disaster are the cost of the solution and the recovery time objective. At the high end of the cost spectrum are the replication and clustering methods designed to provide near-instantaneous recovery at a remote location. At the other end of the cost spectrum are manual recovery methods which cost little in the way of infrastructure but have a much longer recovery time. In general, the cost of downtime drives the design. If downtime is very costly to your business, the solution would move towards the near-instantaneous methods.

By the use of automation of common recovery mechanisms, BMR can markedly reduce downtime without much additional cost. First, automation takes away human intervention and the associated time required for human input and user errors. Second, automation enables a single person to handle more simultaneous recoveries than without automation. Time reduction in recovery of systems without an increase in infrastructure creates the driving force for the BMR solution.

Figure 4.1 illustrates the cost benefits of an automated bare metal recovery solution. Generally, as the recovery time decreases, the cost of the solution increases. Manual rebuilds are inexpensive as a solution, but recovery time can exceed the maximum tolerable by the business. BMR requires no more infrastructure than do manual rebuilds, but the automation has the potential to greatly reduce the recovery time. Hence, automated bare metal recovery solutions fall under the cost curve, making this solution a cost-effective compromise for data recovery.

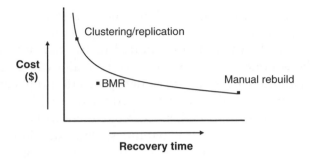

Figure 4.1 Recovery time versus cost

4.2.2 Why Has This Taken So Long?

The principal challenge that all enterprise data protection solutions have faced in performing BMR has been the recovery of the operating system itself. In order to restore the operating system, the backup agent is required, but the backup agent requires an operating system to function. Even more basic, before you can recover the operating system, the data destination such as a file system must be present to hold the restored files.

Creating these file systems for the various operating systems that exist in today's enterprise is a daunting task indeed. Across the spectrum of operating systems, there are many different types of file systems and volume managers that must be recovered, each with different options and capabilities. Each of the different file systems can be created with nondefault options such as disk quotas, large file options, block sizes and so on. Most volume managers also place hidden structures on the disk that cannot be easily backed up or recovered. Even block copies of these structures cannot always be placed onto new disks because the structures themselves must often be defined on disk cylinder boundaries. Also, these structures must be coordinated on all of the disks that comprise a volume or disk group. As an example, a simple block-to-block recovery of the disks in a RAID 5 (redundant array of independent (inexpensive) disks array onto another set of disks will fail to properly recover the RAID volume unless the disks are absolutely identical in number, size and cylinder geometry. Overcoming this rather severe restriction compels the data protection solution to obtain far more insight about the structure of the volume than is normally obtained during a backup. At recovery time, the data protection solution must have the necessary intelligence to reconstruct the volume and create the file system with the proper options and size.

The complexity of saving the volume and file system information for the major operating systems, volume managers and file systems and then providing the capabilities required to use that information to recreate the volumes and file systems is the major cause for the lack of BMR capabilities in the enterprise data protection applications. This gap is beginning to be filled but integrated BMR capabilities have been slow in coming. The solutions being offered today are not yet complete and they vary greatly in their capabilities. Understanding these challenges, limitations and capabilities is important for proper disaster recovery planning and for judging which solution is more capable in your particular environment.

To understand where we are today requires an examination of how we got there. Although the need for an integrated BMR capability was recognized by data protection vendors, the demand was never seen as being great enough to warrant both the initial expense required for research and development and the ongoing expense that would be required to accommodate the inevitable operating system changes that are bound to occur over time. Administrators had to make do with a variety of techniques to fill the gap.

4.3   THE EVOLUTION OF BMR CAPABILITIES

The need for BMR of computer systems has been present as long as there have been systems with unique data. As the number of computer systems increased in the enterprise, data protection methods that provided centralized backup and recovery capabilities became widely used. As these initially lacked the ability to recover systems from bare metal, several methods for fulfilling this need became prevalent. Indeed, most are still in use today. Most of these fall into one of the following categories:

- Manual operating system reinstallation followed by restore of the data from the enterprise data protection solution.
- Recovery of the operating system from an image backup followed by restore of the data from the enterprise data protection solution.
- Hybrid or home-grown solutions locally developed and maintained.

We will look at each in more depth to understand how they lead to the evolution of the integrated BMR solutions beginning to emerge in today's enterprise data protection applications.

4.3.1   The Manual Reinstall-and-Restore Method

Perhaps the most obvious and widely used method for BMR was the manual method of rebuilding the system. This method is theoretically simple:

- Reinstall the operating system.
- Reinstall the applications.
- Install the enterprise backup client.
- Recover the data from the enterprise data protection solution.

The reinstall-and-restore method involves installing the operating system and applications from scratch and then restoring the system's data from its latest backup.

4.3.1.1   Limitations of the manual reinstall-and-restore method

There are several significant problems with this method. We will cover the most important ones here, but keep in mind that these are not the only issues:

- *Inconsistent operating state.* As the system is reinstalled with the base operating system and applications, the user data and other files restored from the enterprise backup may not be consistent with the software that the rebuilt system is running. For example, if a system was running a more recent patch level of an application than what is reinstalled, the application may not function properly with its data files that are restored from the backup. The operating system parameters of application servers are often tuned to maximize the application performance. These changes must be reapplied. There may also be specific user IDs for the application user and group that will have to be recreated for proper application functionality. Application vendors may ship hot fixes for issues not yet released in a patch, or consultants could have provided custom modifications that will need to be reintroduced on the rebuilt system. Even with careful inspection and documentation of the production systems, it is difficult to identify these application and operating system modifications.
- *Overwriting the running operating system.* Once the operating system and the backup application are installed, it is very desirable to recover operating system configuration files such as password and

group files, inittab, host information, system patches and patch history, network services information and so on. In fact, it can be difficult to identify all of the volatile and nonvolatile operating system files that will need to be recovered. When a recovery of this sort is attempted, operating system executables and libraries currently in use could easily be overwritten, and this can cause the system to become unstable. On Windows systems, overwriting the System_State can result in a registry that does not match the system or a hardware abstraction layer (HAL) that does not function on this hardware. This will often lead to an unrecoverable situation that will require starting over again with the operating system installation. Sometimes this instability is not readily apparent until much later in the process when the system is returned to production. This can lead to a tedious trial-and-error process to identify the files that need to be excluded from recovery. Without recovering these operating system files, it can be difficult to get the system back to an operable state.

- *Requires specialized skills.* The reinstall-and-restore process requires specialized operating system skills. Application recovery may also require specialized application skills. In some cases, it is not possible to reinstall the application properly because the installation software is not readily available. System recovery using this method will require people who are intimately familiar with the production environment as well as people who have application and operating system skills. You will not be able to effectively utilize your UNIX administrators to recover Windows systems and vice versa. In a disaster recovery situation, you may not have access to the skilled people in your organization that can rebuild the environment. And even if you are successful in getting the system and application recovered, the final result can be a system that has questionable integrity as it was pieced together from potentially disparate elements.

- *Process maintenance.* Regardless of the solution, to guarantee the recovery process, detailed recovery procedures must be created and maintained. However, the reinstall-and-restore method will require that these procedures be reviewed and tested more often than other methods. For example, they must be updated when operating system changes such as upgrades, application changes and tuneable parameters are changed to ensure that the system can be recovered in the event of a system failure or disaster. This process maintenance places an extra burden on the information technology (IT) staff. Changes are sometimes made due to specific problems after lengthy trouble-shooting sessions in which several parameters may have been changed. These changes are too often done without thought given to the data recovery procedures. In practice, therefore, process maintenance is difficult to enforce.

*   *Lengthy and error-prone recovery.* It can take days to rebuild a system. Rebuilding the system will include reinstalling the operating system and applications, restoring the data, fixing the configuration and reconciling any differences between the reinstalled and restored pieces. In the end, it is often impossible to completely recover the system. In addition there are many opportunities for errors along the way. Many of these errors are subtle and could take additional days to discover, troubleshoot and rectify.

Some systems have installation applications that can potentially reinstall the system with the same level of operating system and applications it had before a disaster. Some examples of these are listed in Table 4.1.

All of these installation applications except Windows also allow operating system patches and application software to be installed. This allows a system to be rebuilt with the same operating system level, the same applications and even the same application patch level. However, this is difficult to achieve in real practice. If all the systems were identical, it would be relatively easy to keep the installation tools in sync. In production, though, systems typically do not remain identical. Client/server applications and business needs will always drive differences into the systems. Installation applications can accommodate system divergence by creating definitions for each class of the system. However, this requires that the installation application be constantly in sync with all of the changes to each system's configuration and software revision levels. If an application has been upgraded or an operating system patch applied, the installation application's corresponding system definition must also be updated. Otherwise, the system will not be able to be reinstalled with the same software it had before the failure. Given

Table 4.1 Installation applications

Operating system	Installation application
AIX	Network install manager
HP-UX	Ignite-UX
Solaris	Jumpstart
Window	Unattended text file
Linux	Kickstart

the nature of system administration in complex environments, it is not surprising that the installation application's view of a system often diverges from the system's actual state.

Another problem with these installation applications is that they are unique to a given operating system and differ from platform to platform. These differences typically mean that valuable administration skills and processes developed for one platform cannot easily be adapted to other platforms. It also means that several different recovery methods must be administered and protected. In this environment, an administrator experienced in a given platform must usually perform the system recovery. Because of the intense nature of system recovery, this significantly limits the number of restorations that can be done simultaneously.

And finally, these methods are slow and inefficient. In the case of a Solaris system that uses jumpstart to rebuild a system and applications to a particular level, it is not uncommon for many of the operating system files to be installed numerous times as the various operating system patches are applied to the system.

4.3.2   Operating System Provided Recovery Solutions

The shortcomings of the reinstall-and-restore method and the as-yet unfulfilled ability of the enterprise data protection solutions to perform BMR created a gap initially filled by the operating system vendors. The operating system vendors were more capable of solving the problem of creating the file systems and recovering the operating system – a task they justifiably saw as a core requirement. Some examples of these are listed in Table 4.2.

The operating system recovery solutions do a good job of recovering the system and its configuration to the exact state that existed when the system image was taken. Most, however, can only perform full backups and may back up the same data as the enterprise data protection solution. Although these operating system solutions provide a

Table 4.2   Recovery tools

Operating system	Recovery tools
AIX	mksysb, sysback
HP-UX	make_recovery
Solaris	Flash archive

means of formatting disks and restoring the operating system, little effort was placed on protecting, cataloguing, storing and retrieving these backup images, not to mention the protection of important user or application data.

4.3.2.1   Limitations of operating system provided recovery solutions

Administrators attempting to recover their systems from bare metal using the operating system provided recovery solutions had to contend with a variety of issues:

- *Specialized skills required.* Operating system provided solutions are unique. They often require specialized skills that cannot be leveraged on other operating systems. Recovery of an AIX system will require a set of skills that differs greatly from that required to recover Solaris or Windows. Distinct islands of expertise are generally acceptable for the recovery of a single server in the data center because the people required for the recovery are readily available. However, multiserver or large-scale data center disasters such as those that occurred in the World Trade Center destruction on September 11, 2001 or the devastation of New Orleans by hurricane Katrina in 2005 have shown clearly that it is not always possible to count on the availability of these key personnel.
- *Wasted storage, bandwidth and effort.* Operating system provided solutions require separate images and lack the robust facilities present in the enterprise data protection solutions to protect this valuable data. These system imaging solutions usually require a full backup so they use a huge amount of storage space and network bandwidth. A typical boot disk backup is at least several hundred megabytes, and a full system backup can be many gigabytes. Multiplied by tens or hundreds of machines, this represents a significant amount of data to back up on a regular basis. These backups can cause performance problems on local networks and take up a large amount of dedicated disk and tape storage. Some of these solutions also require that the system be quiesced or in some cases even shut down in order to obtain a coherent image.

    These system image backups can require a lot of effort as they often require manual initiation by administrators and operators. Administrators and operators must also mediate conflicts between the system image backups and the normal enterprise data backups, gather and label tapes, and verify the success of each backup.

- *Out-of-date backups.* Because system image backups are expensive in terms of downtime, storage, network usage and human effort, they are usually done less frequently than incremental backups. Organizations will often perform image backups of their systems only once a week or even once a month. As a result, when restoring a machine from the system image backup, that backup may be completely out of sync with the data on the enterprise data protection server.
- *Uncertain recovery.* Trying to bring the system up to date with the latest backup is fraught with problems as well. File systems that were created after the image backup will not be recreated because they are not in the image. File systems that were removed after the image backup was taken will be recreated and recovered from the image backup. File systems that were expanded after the image backup was taken are back to their original size. Files and directories that were moved or deleted since the image was taken are back in the locations in which they were originally. Updating the recovered system using the backup software can result in full file systems, out-of-date files and multiple file copies in different locations. Cleaning up the mess afterwards requires knowledge of when files or file systems were created or deleted, moved and renamed. In a full file system situation, there may not be enough room to recover the critical data and the file system may have to be recreated and then re-recovered in order to proceed.

The end result is a system that is most certainly not the same, and the differences can cause problems that may not be known immediately. There is no way to ensure a coherent recovery of the system when you are trying to patch together data that was backed up at different times with different backup methods. The real issue – the source of the inherent difficulty – is that the data protection solution and the operating system provided imaging solution have no knowledge of each other.

4.3.3   Hybrid or Home-Grown Recovery Solutions

With the problems presented above, it is perfectly understandable that companies with a sufficiently large IT staff or budget would develop their own BMR solutions to meet their unique requirements. These 'home-grown' solutions require a great deal of expense and expertise to maintain. The solutions must be evaluated when new versions of the operating system are deployed or when new applications are installed. Documentation of the procedures and processes required must also be maintained. The development staff needs to be available should any difficult or confusing situations arise in disaster recovery.

However, as stated earlier, you cannot always count on the availability of key personnel during a disaster. All the expense of developing in-house solutions could be for naught if the solution cannot be made to work during a major disaster.

Tales From the Real World

Here is an example illustrating the weaknesses of these system recovery practices. A data center takes mksysb system backups of an AIX server every weekend. This machine acts as a server for network information systems (NIS), domain name system (DNS) and some applications.

During a typical workweek, a system administrator did the following:

- *Added two new users to NIS.*
- *Installed a new application.*
- *Added a new application server's IP address to DNS.*
- *Changed the configuration of another application.*

There was a huge storm on Friday, and the roof leaked water on the AIX server damaging it beyond repair. There were reports of some local flooding, but the damage to the building was relatively minor, this system was the only one affected. The company eventually found a replacement machine with a similar configuration. The administrator restored the new machine from the original server's mksysb image, which brought its system configuration back to its state of the previous weekend. The administrator then used NetBackup to restore the server's applications and data, which she hoped would bring the machine back to its state as of Thursday night.

However, all did not go smoothly. Far from being back in business, the administrator now had to grapple with the following problems:

1. *The applications' directory was recovered from the image, but the file system in which it was stored became full when attempting to bring it up to date with Thursday night's backup. It was thought that some files must have been deleted after the mksysb was taken and were therefore restored by the mksysb. There were new files created as the mksysb and these files were backed up with the*

NetBackup incrementals. When the NetBackup incrementals were applied, these filled up the file system during the restore attempt. The application administrator was called for guidance to identify which files were likely the ones to be removed, but the administrator was not available. It took over 2 hours to find files that were the likely culprits, and these files were moved to another file system just in case they were needed. The restore was completed but only one of the applications started successfully. Its inittab entry was found to be missing. This was corrected, but another nuance was found: The newly installed application was not listed in the AIX operating systems application list (lslpp). This did not cause an application error but would have to be corrected so that future updates of the application could be done.

2. *The users' home directories and files were restored successfully from the mksysb, but the newly added users could not access them. The system did not recognize the users because the NIS database regressed to last weekend's state.*

3. *Users were unable to communicate with the new application server as the DNS configuration files no longer contained the server's IP address.*

4. *The application whose configuration was changed had regressed back to its old behaviour. It was discovered that the application's configuration files were excluded from the NetBackup backups. The reason for the exclusion was not clear. But as its configuration files were restored from the out-of-date mksysb, and the new configuration files were not backed up with NetBackup, it was suspected that the regressed behaviour was due to these configuration files. After the configuration files were manually updated, the application was still misbehaving. It was not clear if the problem was due to the improper updates to the configuration files, or some other reason. It was thought that perhaps those files which had been moved to the other file system due to the disk-full errors during restore may be the reason for the misbehaviour. The application administrator was not returning the pages, and his home phone line was down. One of the operators was sent to his home in an attempt to locate him.*

After many hours of puzzling over the system's behaviour and fielding user complaints, the administrator finally pieced together enough of the missing files and configuration to get everything back to a functional state, but there were still issues with the application. The inventory screen was not displaying all of the fields, and updates done on Friday before the outage and after the backup could not be done properly. The application users were seeing performance issues and the application would occasionally 'lock up'. The IT staff suspected that there was some other issue – possibly operating system related – and continued to troubleshoot the application with the application vendor. The application administrator was finally located. The entire ground floor and base-ment of his home were flooded (destroying his pager) and the road to his house had been

(continued)

washed out. His immediate concern was to arrange shelter for his family until the house could be repaired. He was not in a position to help with the application.

And this company was lucky. First, it was able to find its most recent system backup – not always a sure thing given the complexity of juggling two different backup methods and the inevitability of Murphy's law, which predicts that everything that can go wrong will go wrong. Second, the company was using an operating system that had a system backup solution. If it was using an operating system like Linux, the administrator would have had to reinstall the machine from scratch, and then restore data from NetBackup and try to make it all work together. And hopefully the failed server was not being used for critical production purposes. If it was, the company could have been out of business by the time the server was recovered.

As our example shows, even minor system outages can be extremely disruptive to the enterprise. Consider what would happen in the scenario if multiple servers required restoration at the same time. Few organizations have enough talented system administrators to pull off a recovery process like that. And even if they do, every minute counts in those situations, and the common system recovery practices squander precious time and provide little integrity in the process.

4.4   FILLING THE GAP – INTEGRATED BMR

Limitations such as those described above created demand for a more integrated solution. The first solutions were provided by third-party vendors instead of the data protection vendors, but there were different degrees of integration. Those at the lower end of the integration spectrum simply made use of their existing imaging capabilities, and then stored this image as a single unit in the enterprise data protection solution. Although this offered an enormous degree of protection for the image, it did not solve all of the problems that are created when the image backup is separate from the data protection backup.

It was not until the introduction of the Bare Metal Restore™ products by a small company that was based in Austin, Texas, called The Kernel Group (TKG) in April 2000 did a true integrated solution emerge that utilized the ordinary backup data in the data protection solution to recover a system from bare metal.[1]

---

[1]TKG was acquired by Veritas Software in December 2001 and Veritas was acquired by Symantec Software in June 2005. The BMR product has become an integrated option to NetBackup version 6.0.

Most of the operating system supplied imaging solutions captured information about the file systems and volumes so that they could be rebuilt during recovery time. TKG's innovation was to take this concept to the next logical level. After the volumes and file systems were rebuilt, instead of recovering the data from an image, it simply recovered the data from the enterprise data protection solutions (TSM or NetBackup). This method is starting to now appear in other enterprise data protection solutions in a limited fashion. At the time of this writing, none of them have yet achieved the platform and volume manager support of the Bare Metal Restore^{TM} product. It remains unique and worthy of special attention. At this level of integration, the normal backup is used instead of an image. The enterprise data protection solution, therefore, has full visibility and access to the files that would be used to perform a bare metal recovery.

The solutions that integrate poorly must rely on third-party applications to handle the BMR, and thus require interaction between the enterprise data protection solution and the third-party application. This interaction may require different user interfaces on different servers to accomplish the task of BMR.

4.4.1   The Bar Rises

The solution pioneered by TKG had to overcome two major limitations:

- recovery to dissimilar hardware;
- recovery to dissimilar disks.

Even though the recovery to dissimilar disks could be framed as a subset of dissimilar hardware, they are really two distinct situations. The issue of dissimilar hardware recovery arises because the recovered system may not have device support for the hardware onto which it is recovered. Modification of the recovered system is, therefore, often required for successful recovery. This problem is very different from that of dissimilar disks. In dissimilar disk recovery, the problem is that the original file system layouts may have to be altered to fit on the disks available, but the recovered data will not have to be altered.

4.5   THE PROBLEM OF DISSIMILAR DISK RECOVERY

If the disks on the system were replaced with fewer disks or a larger number of smaller disks, the recovery could fail. The real issue has to do with the size of the original file systems themselves, not their contents. This is because the file systems have to be recreated before the data is recovered. If the file systems in their original size will not fit on the available disks, the recreation of the file systems to their original size would fail and the recovery would therefore not be able to proceed. To solve this problem, the original volume information would have to be adjusted somehow before the volumes and file systems are created on the new disks.

The first bare metal solutions saved the disk configuration information, but not in a form that was easy to edit. In some solutions, the configuration information was stored in the backup image itself. During recovery, it was retrieved and then used to recreate the volumes. This needed to change. This can be done in one of the two ways:

- Changing the configuration information about the volumes and files systems before the recovery takes place.
- Adjusting the size of the volumes and file systems to match the new disks during recovery.

4.5.1   Approach 1: Changing the Disk and Volume Configuration Information

Before the disk configuration information could be modified to fit the new disks, knowledge about the new disks would have to be gathered and be available. This approach required that the BMR solutions be able to

1. present the disk configuration information so that it can be changed before the recovery;
2. provide a means to input the new disk configuration;
3. provide a means to recreate the volume and file system layout of the original disk configuration onto the new disk configuration.

This would require a major redesign in the way volumes are created with the existing solutions. The solution would have to accommodate the changes to the volume sizes instead of recreating them the way they were originally created.

4.5.2   Approach 2: Adjusting the Volumes and File Systems During Recovery

This approach has many of the requirements and elements of the first approach, but moves the reconfiguration of the disk layout to the recovering system. This would require that the work be performed in the limited environment of the recovering client rather than the far more capable environment of the bare metal application server.

4.6   THE PROBLEM OF AUTOMATING DISK MAPPING

Although either approach could be engineered into a solution, the utility of the solution will depend on two criteria:

1. Maintain as much automation at recovery time as possible.
2. Accommodate the different operating systems and volume managers that can be used in the enterprise.

For the first criterion, maintaining automation is a key to successful disaster recovery for the reasons we have described earlier. If the amount of disk space available is decreased from the original, the volume layout would need to change to accommodate the smaller disk space available. Volumes could be recreated at a smaller size. Noncritical volumes could be left off the recovery, or mirrors or RAID arrays could be reconfigured as normal volumes. Automating disk mapping to perform these operations would require the bare metal recovery application know the following:

- The mirrors that can be broken or the RAID volume that can be redefined.
- The file systems that can be safely created at a smaller size.
- The file systems that contain noncritical data and that need not be recovered.

The application could make use of user intent data but this volume manager feature is rarely used and is not available on all volume managers. Without a consistent way of obtaining this kind of data on different operating systems, volume managers, and file systems, the best that can be achieved in the way of automation is to provide a user interface on the bare metal application server to at least facilitate the mapping of volumes to the new disks using a common interface. This design would result in additional user intervention.

The second criterion, accommodating different volume management types, is very demanding. For example, in Solaris the most widely used volume manager is Symantec Veritas Volume Manager (VxVM). It is very often used on disk(s) that contain the Solaris operating system. The Solaris boot process requires that the offsets of the root volume containing the kernel be known to the firmware so that it can locate and load the kernel. VxVM requires that the information of the disk locations (offsets and sizes) for the volumes be stored in a private region of the disk so that it could present the volumes to the operating system. This means that the firmware view of where the root file system starts and volume manager view must be in sync. The solution which VxVM has used is to simply define the root volumes' location information to be that of the underlying disk slice. The result is that the firmware information does not have to be changed. The process by which the disk slice information containing the file system is known is volume encapsulation. In dissimilar disk restore, if the size and physical location of encapsulated root volumes are changed, both the disk slice offset and size information and its corresponding volume definition for Veritas Volume Manager must be kept in sync. If this does not happen, the system will not be able to start even though all of the data is recovered.

The native Microsoft Windows operating system volume manager is the Logical Disk Manager (LDM). Besides basic disk support, LDM also has support for *dynamic disks*. Dynamic disks can contain RAID volumes, mirrored volumes, spanned volumes and striped volumes. Not to be confused with RAID adapters where this configuration is performed on the firmware of the mass storage adapter, these are defined within the Windows operating system. Recreation of dynamic disks requires the use of unpublished Microsoft APIs.

These solutions are beginning to emerge. A good example of this kind of capability is that provided in the Symantec Veritas NetBackup Bare Metal Restore™ product. Although this product is nearly alone in its support for most of the more widely used volume managers and file systems, we can expect that other BMR solutions will catch up. This solution provides a means to discover new disk information and provides a common user interface to map the original volumes to the new disks. It requires user intervention to perform this mapping, but the intervention can be done in a common user interface for all platforms and supported volume managers ahead of the actual recover itself, maintaining automation at recovery time.

4.7   THE PROBLEM OF DISSIMILAR SYSTEM RECOVERY

Recovering the system onto hardware that differs from the original requires that device support for the new hardware be present on the recovered system. If it was not installed on the original system when it was backed up, the recovered system will have to be modified so that it can run on the new hardware. The challenge is that the recovery environment has limited capabilities. The device installation usually requires specific operating system commands, utilities and APIs. BMR solutions that utilize alternate operating systems to perform the recovery will have issues at this stage.

The operating system also needs a means of installing device support into an alternate root directory. This capability exists in the UNIX and Linux operating systems but does not exist in the Microsoft Windows operating systems. In fact, recovery to dissimilar hardware in UNIX can be performed manually after the BMR solution has recovered the system onto the new hardware. If the system will boot, device support for missing hardware can be easily added. If it cannot boot, an alternate copy of the operating system can be booted from the network or from the operating system installation media. Then the recovered root file systems can be mounted onto an alternate mount point and boot device support can be added into the alternate root using standard operating system installation tools.

Recovery to dissimilar hardware is far more difficult to perform with the Windows operating system. First, there are many third-party devices that are used on Windows systems. Device drivers are supplied by different vendors and many are not a part of the operating system. Also, it is very difficult to install device drivers for hardware that is not present on the system. If device support for the boot disk is not installed, the system will blue screen with an inaccessible boot device message. The method that Microsoft supplies to install new boot device support into the system (sysprep) requires that the system be prepared with the new disk drivers before the backup is performed. This method is generally not effective in disaster recovery scenarios where you cannot always know the hardware onto which you will recover before the backup is done. It also does not accommodate HAL changes.

4.7.1   Windows Dissimilar System Restore Issues

Recovering Windows systems to different hardware is a daunting but necessary task. When a Windows system fails and needs to be replaced, it is not always possible to obtain a replacement system with the identical hardware. The replacement hardware can differ in a number of ways:

- mass storage drivers
- HAL
- speed and number of CPUs
- network interface cards

Mass storage drivers. The first issue that needs to be solved is the disk driver issue. Without being able to read and write the operating system data on the disk, it is not possible to proceed further. Even if the disk driver is supplied by the operating system, it must be started very early in the boot process so that the system can load the operating system. The BMR solutions that support recovery to different hardware have in common their ability to provide OEM disk drivers needed to access the disks onto which the data will be recovered.

Recovering the data onto the new disks is just a part of the problem. Lacking any operating system supplied facilities, the BMR solution must itself provide a means to install the disk driver files on the recovered disk in the appropriate location and must modify the recovered registry to start the inserted driver at boot time.

HAL. The next major obstacle to dissimilar system restore of the Windows operating system is that of installing the correct HAL. The HAL provides a means to translate the common Windows operating system APIs to hardware specific routines. There is a limited number of HALs, almost all of which are supplied by native Windows installation tools. Once the appropriate HAL has been identified during the original Windows installation, it is renamed and placed onto the system as HAL.dll.

During recovery, the HAL.dll may not be appropriate for the new hardware. If it is not correct, the system will not function properly on the new hardware and may fail to boot. The BMR solution must be capable of identifying and installing the appropriate HAL onto the system. The HAL.dll file is sometimes updated by Windows operating system updates. This complicates the installation of the proper HAL onto the recovered operating system because a HAL update must be accompanied by other operating system file updates. If the HAL is not at the same patch level as the operating system files on which it depends, the system can become unstable and may blue screen.

Speed and number of CPUs. If the original system had a single CPU and the new system has multiple CPUs, the kernel must be configured to run on multiple CPUs or the system will not utilize the additional CPU(s). The BMR solution must be able to make these changes.

Network interface cards. The network interface cards may be different on the new hardware and the original static IP addresses must be bound to the new interfaces. This is not too difficult if there is only one interface but if there are multiple interfaces the BMR solution must provide a means of mapping the original IP addresses to the new interfaces. If the network interface is not correctly configured, it will be difficult to perform a restore from the network. Setting the IP addresses manually is an alternative that can be done at the expense of automation.

Recovering network teaming configurations can also be difficult if the underlying network interface cards are different. Network teaming can also be done manually, again at the expense of automation.

4.7.2   UNIX Dissimilar System Restore Issues

The major UNIX systems are built on proprietary hardware, so the issues of recovering to dissimilar hardware on UNIX are more easily overcome. There are very few UNIX system devices that do not have drivers in the base operating system distribution. AIX and Solaris usually do not install device support for all the hardware available so it is very possible that critical device support for the new hardware could be missing on the original system. This can cause a recovered system not use the new hardware. On the contrary, HP-UX usually installs nearly all device support along with the operating system. Recovering HP-UX systems to different hardware, therefore, is usually not an issue.

It is possible to overcome the issues in the following ways:

1. Install support for all known devices on the system before it is backed up. Unlike Windows, all UNIX systems allow support to be installed for devices that are not physically present. That makes this a straightforward and easy task to accomplish. For Solaris, you can do a full OEM installation of the operating system. For AIX, you can install all devices at any time.
2. If that is not possible, you can install missing device support onto a system that has been recovered but fails to boot. This is also not difficult. The system can be booted from the operating system installation disk and the root file system mounted in a temporary location. Then the missing device file sets can be installed into the mounted root file system.

Surprisingly, data protection integrated BMR solutions that automate recovery of UNIX systems to dissimilar hardware are not yet available. We are currently left with having to perform these operations manually.

4.8   THE CURRENT STATE OF INTEGRATED BMR

This takes us up to the present where we can look at the kinds of innovations that are available today. This is a rapidly changing field and the information here might be out of date by the time this book is published. There are several BMR solutions available today that provide varying degrees of integration and provide varying degrees of automation.

There are a few solutions that provide integration by storing third-party images in the data protection solution. These have adapted a rather novel approach to providing a coherent system. After the third-party image is used to recover the system, the system files are brought up to date with the latest backup from the enterprise data protection solution. This is a special invocation of the recovery that will delete files that did not exist as of the time of the backup and restore the file permissions on the files as they existed at the time of the backup. However, it is worth repeating that enterprise data protection solutions cannot resize file systems that were resized or recreate file systems that were created after the third-party image was taken.

Perhaps the most noteworthy innovation that has been developed was the client configuration concept that is used in Symantec Veritas Bare Metal Restore™. This concept is worth noting because it provides a means of automating recovery and lays the basis for far more automated recovery in the future. The concept is simple. The client configuration (known simply as the 'config') can be thought of as an abstraction of the system. It is stored as an entity in the application database. The client's config is maintained at backup time and is named 'current'. The current config can be copied and the copy can be extensively edited using a unique configuration editor in the Symantec Veritas NetBackup™ Administrative GUI. Windows disk and network drivers, client IP addresses, network routes, NetBackup Client configurations and disk volumes can be changed extensively using the configuration editor.

As the configurations are stored as independent entities in the database, the original client does not have to be available for the editing to occur. The client configurations are also saved in the backup data for each system and can be retrieved with the configuration editor from NetBackup to perform a point in time restore.

This concept of a client config is indeed powerful and innovative. For example it means that the administrator can decide onto which hardware the client will be recovered *after* the client suffers a catastrophic failure. It means that all of the changes required to restore the system onto new hardware can be done using a common interface in advance of the restore. This allows the restore itself to be as automated as possible, requiring only minimal if any manual intervention. This design lends itself extremely well to the pressures that administrators experience during system recovery and allows a single administrator to recover numerous systems simultaneously.

4.9   THE FUTURE OF BMR

4.9.1   Enterprise Data Protection Server Self-Restore

Perhaps the largest hole in the bare metal recovery technology today is the bare metal recovery of the enterprise data protection server itself. It is always good at this point in the book to provide a humorous analogy that illustrates the problem: A bare metal recovery of the data protection server is like a brain surgeon operating on his own brain. The surgeon removes his (or her) brain and places it on the operating table then stands there in front of it, incapable of making the slightest decision on how to proceed.

All of the enterprise data protection solutions today are only capable of recovering files if the data protection solution is functional. In a bare metal state, without the brains of the operation available, the server will sit there, incapable of recovering anything.

The enterprise DP solutions can store data on a wide variety of tape and disk devices. They have catalogued which tapes on which libraries contain which data. The catalogue information for the files of the server itself must be known before the recovery of the server can take place, and device support and connectivity to the devices must be in place before there is hope of recovering any data. This is the challenge that must be met and overcome.

The good news is that all of enterprise data protection solutions are capable of being recovered from bare metal. The bad news is that none of them are automated, and all of them require specific sets of skills to perform the recovery, skills to rebuild the operating system and skills to install and configure the enterprise DP server system.

Work is being done in this area to automate the self-recovery of the enterprise DP servers, but we are not likely to see any results of this until late 2007.

4.9.2   Automated Dissimilar Disk Restore

The automatic mapping of disks layouts to alternate disks is the focus of research for obvious reasons. Currently, this is a manual process. The difficulties in automation are largely due to not having enough information to automate the decisions required to recover to a smaller number of disks or disks of lower capacity. New information provided by IT administrators about volume constraints or user intents will have to be made available to the bare metal application to solve this issue. The end result will be very useful.

4.9.3   Automated Dissimilar System Recovery

The automation of Windows dissimilar system recovery will depend on the ability of the bare metal recovery solutions to discover information about the new hardware and be able to leverage that information to automate OEM driver selection, IP address mapping, HAL and kernel selection, and disk and volume mapping.

The automation of UNIX dissimilar system restore is long overdue and should be easier to achieve.

4.9.4 Network Integration

Bare metal recovery of the network infrastructure is also an unfulfilled need that requires attention. Storing and being able to edit network settings in coordination with system configurations would provide tremendous value that could automate switch and router settings to create gateways and to configure VLAN and port settings. This will help large-scale bare metal recovery tremendously.

4.10   NEW CAPABILITIES AND CHALLENGES IN DATA PROTECTION AND THE EFFECT ON BARE METAL RECOVERY

4.10.1   Continuous Data Protection (CDP)

CDP technology will enable more granular point in time bare metal recovery capabilities. One of the more overlooked areas in CDP is to provide help for the user in deciding the recovery point. Nearly all of the CDP solutions provide the ability to bring a file back from a point in time, but they lack the ability to help the user decide which point in time to use. For example, if you had a corruption of a set of files in a directory folder, but did not know exactly when the corruption occurred, how would you know to which point in time to recover? To add any meaningful capability of CDP technology, event awareness needs to be brought into the picture.

4.10.2   Single Instance Store (SIS)

SIS technology can provide a tremendous value to bare metal recovery. This will enable remote office recovery and efficient bandwidth utilization capabilities that simply do not exist today. If you need to recover 50 Windows servers at a remote office, nearly all of the operating system files will be common across these 50 systems. SIS can reduce the amount of data that must be sent across the wide area network (WAN) link by a huge amount, allowing a remote office recovery from centralized data to be feasible.

4.10.3   Storage Area Network (SAN)

Automated and coordinated provisioning of SANs with an intelligent bare metal recovery capability can bestow a powerful means of ensuring that systems have enough storage capacity to be recovered. The flexibility that coordination of these technologies could provide would be incredible.

4.11   LARGE-SCALE AUTOMATED BARE METAL RECOVERY

Imagine showing up at a disaster recovery site with a set of system data from several hundred systems – perhaps on disk or maybe on tape. You have a set of hardware available to you, but know little about how much disk space is available, how the SAN is configured. You identify the system that will become the enterprise DP system, make the systems data available to it (place the tapes in the library or give it access to the data stored on disk). You boot it from a special DVD that was sent with the systems data. Within minutes it recovers the enterprise DP server. Now you initiate a GUI console, select the systems you need to recover and issue a single command. The network routers and switches are located and automatically configured for the discovery phase. You then perform network boots on all the hardware available to you. As each system boots, information about its disks drives, disk adapters and network hardware is discovered and sent back to the enterprise DP server. The server automatically allocates systems to be recovered to available hardware, bringing back the AD servers and DNS servers first, creating and allocating LUNs from the SAN as needed for the systems. The network routers and switches are reconfigured as needed, and within a few hours, all of the several hundred systems have been recovered, and a report is available that summarizes the process. Details are available for each system in the master log.

It will take years to get to this point, but the basic technology needed to achieve it is already here. The future looks bright indeed.

4.12   SUMMARY

In this chapter, we have taken a look at the emerging technology of BMR. We looked at the background of the problem itself and then at some of the challenges that have made the development of BMR difficult. As we have seen, business needs have driven enterprise data protection vendors to provide BMR solutions as a part of the core product. We can expect that this trend of integration will continue to evolve rapidly, and we will continue to see even more innovations in this area. We see the config concept introduced by Veritas as something that will continue to evolve. The next stage is the evolution of tools that will automate editing of these configurations, increasing the level of automation for recovery.

Chapter 5

Management

5.1   INTRODUCTION

Data accumulation outpaces information technology (IT) budgets in
almost all organizations. This is a fact that IT professionals have been
forced to contend with for many years. In an effort to combat this
pressure, administrators have demanded technology innovations that
would give them greater visibility and control in the data protection
environment. An ideal storage and data protection infrastructure and
complementary tool set should support the backup administrator's
efforts to perform the following tasks more effectively and efficiently:
(1) provide foundational data recoverability for all business data; (2)
consistently measure and meet service level agreements (SLAs);
(3) accurately and efficiently allocate costs back to business units; (4)
meet regulatory compliance requirements.

Provide foundational data recoverability for all business data. In
many organizations, servers and storage resources are deployed by
multiple groups, and in many cases deployments are done without
inclusion of data protection. This challenge needs to be addressed on
two fronts. First, the application and file server deployment process
needs to be clearly defined, strictly enforced and must include data
protection as a key component. Ideally, each resource is protected to
recovery point objective (RPO), recovery time objective (RTO) and
backup window levels commensurate with the business value of the
resource. A baseline goal is to ensure that no server is deployed without
minimal data protection which in most organizations is nightly incremental tape backups and periodic full backups.

Digital Data Integrity   David Little, Skip Farmer and Oussama El-Hilali

Consistently measure and meet SLAs. IT organizations are now commonly measured by SLAs which establish target objectives for quality and timeliness of services provided by the group. For example, the team might be expected to recover any mission critical server back to production within 1 hour (RTO) of a server crash or data loss incident and lose no more than 1 hour worth of production data in the process (RPO). In order to manage an operation to this level of efficiency, first, adequate management tools need to be implemented in order to have an ability to achieve the objective of the target goals. Then adequate monitoring and reporting tools need to be implemented in order to quantitatively measure the team's performance in meeting the defined objectives.

Accurately and efficiently allocate costs back to business units. IT expenditures have become a major component of the overall organizational budget for many companies. Although IT is commonly a centralized function within companies, organizations increasingly desire the ability to allocate costs back to functional business units, which provides two primary benefits: (1) it can support an accurate reflection of the true discreet costs associated with a particular function within the organization and (2) this approach enforces a shared accountability to IT resources so that business owners can partake more fairly in the negotiations of trade-offs between costs and SLA delivery.

Meet regulatory compliance requirements. For many organizations, compliance requirements affect numerous aspects of business, from the back office to the production floor to the field sales organization. Regulations such as Sarbanes-Oxley, 17a-4, NASD 3010/3011, FDA Part 11, HIPAA and so on have particular interest in information assets within an organization. Rules vary from regulation to regulation and industry to industry, so it is imperative to seek legal counsel and to clearly map the specific regulatory requirements that govern a particular organization, and to establish, document and enforce processes that support compliance to these regulations. In examining the applicable regulations and their bearing on a data protection environment, some common themes can be found. In addition to specific requirements enforced by each regulatory agency, most governing bodies will expect an organization to

- produce documented, consistent processes for data protection;
- maintain the ability to manage governed data over a defined life cycle and on an appropriate storage medium as dictated by a specific regulation;

- demonstrate and regularly test the recoverability of backup and archive data;
- provide security for sensitive data as it remains within the data center as well as when data is transferred off-site for disaster recovery purposes.

5.2 PROTECTING DATA THROUGHOUT ITS LIFE CYCLE

Although there are numerous types of data within an organization, and each type can be considered to have its own distinctive life cycle, at its core, data goes through some fundamentally similar phases, namely

Creation. In the creation phase, users enter data into a centralized system or author new content on their desktop computing environment. In the creation phase, the storage architecture needs to be designed for fast response time for optimal user experience, and data protection needs to be implemented to quickly capture the first appearance of data, whether on a file server, application server or desktop.

Use/process. In the primary use phase, data is processed according to a formal workflow, as in an ERP or workflow processing system, or accessed in an ad hoc fashion as in Internet/intranet, email distribution, content or document management, network file server share and so on. In this phase, a storage architecture needs to be responsive enough for the demands of a particular application, but not over-engineered to deliver maximum performance where it is not needed, as maximum performance equates to higher costs in storage architectures.

Reuse/repurpose. In this phase, data is accessed and either used again for its original intended purpose, such as accessing a logo and placing in a new brochure, or used for a purpose not originally intended at the creation of the data, such as extracting content from a PowerPoint slide and using it for a white paper. For this phase, the architecture needs to accommodate flexible search capabilities coupled with relatively quick access to data. From a protection standpoint, the system should be able to quickly back up the first appearance of new or modified data so that the window of vulnerability to data loss is quickly closed.

Archive. In the archive phase, data is retained only for retrospective purposes. The guidelines for discovery may be determined by internal corporate mandate or defined best practices, or by external pressures from regulatory agencies such as the Securities and Exchange Commission

(SEC), Food and Drug Administration (FDA), National Association of Securities Dealers (NASD) and so on. In this phase, performance is not critical, but the architecture needs to provide low cost, security, longevity as mandated by policy, reliable access during the retention period and full expungement at the end of the life cycle.

Discovery. As a common thread for data across its life cycle, a storage architecture needs to accommodate an ability to easily search and retrieve data at any point from creation to archive. This has been often overlooked in the past, but recent regulatory pressures have led organizations to reengineer their storage environment to accommodate this requirement. This requirement necessitates the ability to search for data using key words and has an ability to access the data reliably using intuitive means, ideally with minimal intervention from system administrators, as this can become a costly endeavour.

When planning to architect or re-architect a storage environment for optimal manageability and protection, it is helpful to consider the requirements of users, as well as the organization's expectations, layered against the interests of any governing body which has jurisdiction over the organization and its data. This will help guide decisions made relative to storage architecture, migration technology, management tools and data protection infrastructure designed to meet the unique RPOs and RTOs for each type of data at each phase in its life cycle.

As a prerequisite for architecting an environment which is manageable by its nature, it is critical to understand the value of various types of data within the organization and how that value changes as data ages. An important element to this task is to understand that there are two distinctly different measures of value associated with data. The first value can be measured in terms of the value of that data to the business itself. This can be challenging to gauge, but the easiest way to measure this is to consider all of the ramifications of losing this data. In some cases, such as data contained within an ERP system, if the data is lost permanently, then statistically the organization itself has low chances for survival, so the data itself has extremely high value. On the other end of the spectrum, you may have pictures of an employee making unprofessional gestures at a recent company event, which have little if any value to the business and are more than likely considered a waste of resources from an IT perspective. As these two pieces of information age, the value of the ERP data gradually declines from a business value standpoint, whereas the business value of the pictures was low to begin with and certainly did not gain value as it aged.

The other type of value of data that needs to be considered is based on the requirement to reproduce the data for reasons related to regulatory compliance or legal discovery. In this case, the value of data remains somewhat constant over the period during which it is subject to discovery. Another interesting characteristic of the discovery value of data is the abrupt decline in value as it reaches the end of the period during which it is required to be discoverable. At the end of this period, most organizations immediately consider the data a liability, rather than an asset, and it becomes a high priority to remove the data and associated metadata from all storage: online, nearline and offline.

In the previous example of ERP data and the photos, it was obvious that each of these data types has a drastically different value to the business. If one is lost, the viability of the business itself may be jeopardized. If the other is lost, little or no business impact will be incurred. That is where the business value of data can differ from the evidentiary value of data. Consider the possibility that the employee in the photos has been fired and has filed a suit for wrongful termination. In order to defend itself, the organization needs to produce evidence to support its position that the firing was justified. In this case, the discovery value of the photographs can be considered quite high, in spite of the low business value placed on such data. The evidentiary value of this data can be further bolstered by demonstrating that it was produced in its original form without modification, from a system designed to manage and protect data throughout its life cycle continuum.

5.3 ARCHITECTING FOR EFFICIENT MANAGEMENT

As an IT environment grows, it can become unwieldy. Storage architectures become challenged to scale with the demands of rapid data growth, and it is unrealistic to expect any product to layer on top of a chaotic environment and bring full relief to the situation. A preferred approach is to periodically re-architect the storage environment, both primary and secondary, to allow data protection SLAs to be maintained at current and planned data levels. For example, it was common practice in the early 1990s for organizations to procure a server and add backup software and a tape drive as an 'add-on' to the server. This approach led to a highly fragmented environment which yielded high administrative and operational costs for managing data protection operations as the environment grew. Many organizations learned that by standardizing on a single backup product with a client/server architecture using local area network (LAN) backup, tremendous efficiencies could be gained.

In order to re-architect a storage infrastructure that leverages modern technologies that can assist in extending manageability within the environment, it is recommended that an organization go through a meticulous analysis of data types, values, business expectations and recovery requirements for each type of data. Some of the steps to be undertaken in this analysis include data inventory, storage consolidation, tiered storage, data classification and standardization. Let us take a deeper look at each of these steps.

Data inventory. The first step in any project to re-architect a storage environment is to establish an understanding of what data is stored within. First, raw metadata should be collected describing all of the data stores within the organization. In the past, this would have been an arduous task of performing a directory listing on each file and application server, and compiling the information in a central location. Today's SRM technology delivers automation to this task, allowing for file system agents to collect this data with relative ease and centralize it in a way that makes analysis much simpler. In some cases, the SRM technologies provide the ability to perform the necessary analysis within the tool itself (see Figure 5.1).

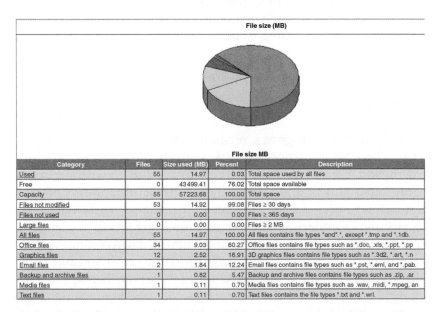

File size MB

Category	Files	Size used (MB)	Percent	Description
Used	55	14.97	0.03	Total space used by all files
Free	0	43 499.41	76.02	Total space available
Capacity	55	57 223.68	100.00	Total space
Files not modified	53	14.92	99.08	Files ≥ 30 days
Files not used	0	0.00	0.00	Files ≥ 365 days
Large files	0	0.00	0.00	Files ≥ 2 MB
All files	55	14.97	100.00	All files contains file types *and*.*, except *.tmp and *.1db.
Office files	34	9.03	60.27	Office files contains file types such as *.doc, .xls, *.ppt, *.pp
Graphics files	12	2.52	16.91	3D graphics files contains file types such as *.3d2, *.art, *.n
Email files	2	1.84	12.24	Email files contains file types such as *.pst, *.eml, and *.pab.
Backup and archive files	1	0.82	5.47	Backup and archive files contains file types such as .zip, .ar
Media files	1	0.11	0.70	Media files contains file types such as .wav, .midi, *.mpeg, an
Text files	1	0.11	0.70	Text files contains the file types *.txt and *.wrl.

Figure 5.1   File system analysis performed by the storage resource management tool Reproduced by permission of Symantec Corporation.

Some of the information that is important to understand includes

- *File types.* A thorough understanding of what file types exist within the data center can help determine the business value of data on average, which provides some of the background information necessary to architect for optimal user experience and data protection.
- *Redundancy.* It can be beneficial to assess the level of redundancy that exists for each file type within the organization. In some cases, redundancy is necessary to provide higher level of availability or to provide multiple access points to the same information to balance I/O across multiple resources to avoid bottlenecks. In other cases, redundancy is unnecessary and simply wasteful. For every copy of a piece of data on primary storage, there may be mirrored and replica copies on disk and numerous backup copies on tape. Therefore, the elimination of redundant files on the primary disk is a high-leverage solution for streamlining storage and data protection operations.
- *Age and access patterns.* As discussed previously, business value can often be measured as a function of the type of data, offset by the time that has lapsed since its most recent update. Another measure, related to these but distinctly different, is the amount of time that has lapsed since the data was last accessed. Some data has a short shelf life. For example, a standard daily inventory report might be quite actively accessed during the day it is generated, but is typically not accessed much after the next day's version is produced. Other data, such as a product manual or annual report, may have relatively higher levels of access over longer periods of time. By understanding these patterns, data can be classified and stored on equipment delivering appropriate levels of uptime and throughput performance commensurate to the business value of the data.

Storage consolidation. By moving from a fragmented storage environment characterized by direct attached storage on decentralized application servers to a consolidated architecture, tremendous economies can be achieved. In addition to producing an environment that is more conducive to general storage management workflow, including provisioning, allocation and so on, this environment is also easier and less costly to protect. Storage consolidation can be done through physical or virtual means. By physically consolidating into a dense, high throughput configuration such as a fibre channel based storage area network (SAN), significant performance gains can be achieved for application and file serving, while also adding value to the backup and recovery operations. Virtual consolidation can be added by implementing a storage management layer as a means of extracting the storage management intelligence from the disk arrays and centralizing it as a standard front end for accessing any back-end storage. This approach is called storage virtualization and can serve to standardize the administration of multiple storage devices and establishes a solid foundation for implementing tiered storage.

Tiered storage. As data was categorized in different tiers in an earlier discussion, so now we can apply the same logic to a storage architecture in an effort to lower our overall costs of managing and protecting data. As a part of a storage consolidation effort, or as a refinement of an existing consolidated architecture, multiple storage tiers can be deployed to match the business value of data to the performance and protection requirements appropriate for data of such value. For example, mission critical ERP data can be hosted on high-performance, highly available primary storage and protected using mirroring, snapshots and tape, used primarily for archive purposes. For less critical data, perhaps slower SATA (serial advanced technology attachment) storage may be configured, coupled with a daily tape backup delivering slower but reliable recovery. In recent years, many organizations have ventured to implement tiered storage and have been able to lower the cost of managing and protecting their information assets as a result of this effort. In some environments, technology is introduced in the form of hierarchical storage management or intelligent volume management software that can automatically migrate data across storage tiers as the business value of data declines based on attributes such as file date and last access. This level of automation may be beneficial to an organization if data volumes and sporadic access patterns are the norm. With or without automation, effective tiering of storage can lead to a more manageable environment and allow for more focused utilization of data protection resources.

Data recovery classification. Once the business value of data has been established and a tiered storage design has been created, data recovery SLAs can be agreed upon for each tier. This is a critical intersection between business units and IT, where costs of recovery need to be assessed and accountability fairly given to business units based on the SLAs established for recovery point, recovery time and retention. Once the metrics are established and agreed upon, data protection monitoring and reporting (discussed later in this chapter) become critical to assess how well IT is delivering on its promise to perform and how much IT resource is being consumed by each business unit.

Standardization. A lack of standardization in any aspect of storage and data protection architectures can lead to higher than necessary costs for administration, storage hardware, training and support. This problem is exacerbated by company merger and acquisition activity which often results in further fragmentation of an environment. A recent trend towards technology standardization and vendor consolidation has allowed some companies to reap tremendous cost savings. A successful strategy for standardization is to invest in a software stack that provides management and administrative tools that operate consistently across server and storage platforms. This allows the organization to focus its training investment on a single suite of tools across multiple platforms which eliminates 'knowledge silos' and costly cross training efforts. By consolidating the 'storage intelligence' into a standard software stack, the organization can also lower its server and storage costs as hardware becomes increasingly commoditized and vendors are forced to more aggressively compete for business, whereas the company may have previously been 'locked in' to a particularly vendor due to the high training costs associated with changing platforms. Through standardization in the storage layer, data protection becomes more efficient, as greater consistency exists between the data protection environment, applications and storage, allowing for a more consistent, predictable delivery of service.

Once the data center is optimized for data protection efficiency and a fundamental ability to reliably protect data has been established, additional technologies can be deployed to provide greater effectiveness. Today, there are a number of target areas for focus that can improve the overall manageability of a backup environment after the previous steps have been undertaken. Some of these areas include disk-based data protection, encryption, end user restore and remote office protection. Let us take a deeper look at each of these areas.

Disk-based data protection. For the first 50 years or so, data protection was primarily considered a 'tape thing'. Disk has always had attractive attributes for data protection, but until recently it has been prohibitively expensive to be used for backup purposes. From the standpoint of data protection management, the most desirable attributes of disk as a backup medium in the data center are its capabilities for error handling, which leads to a more reliable and manageable operation. Disk also provides the ability to deliver superior recovery times for individual files or any subset of an entire backup. As data protection extends to remote office environments, disk also plays an important role as it lays a solid foundation for automated data protection and centralized administration. This can eliminate the need for tape management at remote offices, which has proven to be unreliable. Disk also serves as an enabling technology for things such as single instance store (SIS) and end user recovery.

Encryption. A key element to managing and protecting data is the practice of keeping it out of the hands of unintended parties. Several highly publicized incidents of sensitive data finding its way out of the secure control of organizations have fuelled particular interest in an ability to encrypt data as a means of securing data when all other security control measures fail. Depending on the needs and the risk propensity of a given organization, it may be prudent to encrypt data at all points within the computing environment, including at rest, in transit and in its archival state. Alternatively, some organizations choose to minimize cost and disruption by encrypting only data that leaves the confines of the facility, generally in the form of off-site data replication or tape rotation.

Architecting a SIS solution requires careful analysis prior to implementation. A couple of significant issues can cause challenges in operation if not taken into consideration during design. First, when data is encrypted, a key is generated that holds the instructions on how to reassemble the data into its native format. For security reasons, these keys need to be maintained separately from the data itself, which can require an additional administrative responsibility. This process must be meticulously attended to, because if the correct corresponding keys are not available at the time of restore, the backup data set can be considered worthless because if there was a way to recover the data without the encryption key, then the data was never truly secure.

Another design item that must be considered is where the encryption processing will take place. This must be considered because of two primary reasons. First, the encryption process introduces overhead and can cause a bottleneck in the data protection operation. Therefore, the data protection architect should place this burden where it can be most easily afforded. Some of the options may include the following: (a) at the data host; (b) on the backup server; (c) on an appliance in the data path: (d) on the tape drive. Another consideration is that once data is encrypted, it generally cannot be effectively compressed. Therefore, either data should be compressed prior to encryption or only selected data should be encrypted (perhaps only those copies that leave the site), rather than a wholesale encryption of all data.

End user restore. Data protection administrators commonly deal with several types of data recovery operations. A recovery process may be initiated due to a failed disk array, in which case the entire volume is restored from secondary or tertiary storage. In other cases, some portion of data may have become corrupt, and select database tables or volumes are carefully rolled back to a point in time prior to when the corruption first occurred. Other cases involve a server failure or migration to new server hardware, which requires careful reconstruction of the server's kernel to ensure application coherence and compatibility with existing data and system components. All of these operations require careful handling of a skilled data protection administrator. Many other recovery operations require an administrator's involvement to recover files or email messages that were errantly deleted by a user. Although such requests can be disruptive and taxing on an IT operation, failure to respond in a timely manner can result in frustrated users, and in many cases this frustration can extend to customers who are affected by the delay. Fortunately, backup and archive products are beginning to offer tools for end users to facilitate recovery of their own data, which can be done in a timely manner and without impacting system administrators. This can dramatically lower the costs of data protection, while simultaneously allowing the IT organization to deliver a superior level of service on its core data recovery tasks.

Remote office protection. Many organizations have invested considerable resources in people, process and technology to meet the demanding data protection expectations within a data center. With rare exception, however, these same organizations have felt challenged to extend this level of service to data residing in remote offices, outside of the physical reach of data protection professionals. Protecting data in remote offices using technologies designed for a data center leads to two primary challenges. Tape has been the de facto standard backup storage medium for the past five decades. Tape has several key attributes that make it an optimal storage medium for data protection, including relatively low cost, portability, low power consumption and heat generation, and impressive performance for sequential transfer of large blocks of data. Unfortunately, it often delivers lacklustre results when deployed in remote offices, due to the lack of trained IT professionals at many remote offices, which leads organizations to delegate this responsibility to less technically capable staff. This approach can introduce levels of risk that would be defined as unacceptable in most organizations.

Replication technologies address this challenge by bringing copies of this data back to the data center environment where it can be incorporated into the backup operations within the data center. This approach provides a dramatic improvement for many organizations, but a large percentage of organizations have chosen to absorb the risk associated with inefficient backups in the remote office, rather than investing in the bandwidth necessary to copy all of the data back to the data center.

Fortunately, a new technology category has emerged which addresses the remote office backup challenges head-on. The category can be referred to as 'optimized remote office backup', and products within this category utilize SIS (described in Chapter 8) to filter out redundant data at a file, block or other sub-file level for maximum efficiency, and then transfer only the unique data across a wide area connection. By maintaining both a backup data repository at the remote office and a replica copy at the central site, localized recovery can be achieved without consuming expensive bandwidth, whereas data can also be protected off-site for disaster recovery purposes.

By eliminating tape management at the remote site and providing centralized management and administration of a physically decentralized environment, optimized remote office data protection technology provides tremendous gains in administrative productivity and backup reliability. This approach serves to greatly mitigate the risk associated with remote office data protection. It also enhances an organization's ability to enforce policies pertaining to backup success, retention and archive, which supports the overall organizational initiative towards regulatory compliance.

5.4   REPORTING

As IT environments have grown dramatically in recent years, visibility into what is happening within the storage architecture has become clouded. These environments have become extremely complex, with combinations of file servers, application servers, SANs, Network Attached Storage (NAS), heterogeneous servers and numerous operating systems. Add to that the technologies specific to data protection, such as backup software, tape SANs, tape drives and robotics, and you have an environment that can be difficult to navigate.

In spite of this complexity and in spite of the unbridled data growth, organizations need to maintain visibility into storage and data protection. Even in smaller IT shops this responsibility challenges conventional methods of monitoring and reporting, which have largely been an effort of manual data collection coupled with spreadsheet-based reports. The manual nature of this process leads to inaccuracy and inconsistency, as well as an inability to provide real-time information. These forces have propped up a cottage industry based on management and reporting tools for data protection operations, which has led to the introduction of many products designed to assist with the planning and execution of backup and recovery.

There are two different stakeholder groups with distinctly unique requirements for reporting on the backup operation: (1) the IT group tasked with responsibility for delivering backup and recovery as a service to the organization and (2) the business units within the organization that are being served by the IT organization and need to have assurance that their interests are being met by IT. Let us review the specific needs of each of these groups.

5.4.1   Backup Operations Reporting

The backup operations team is tasked with delivering reliable data protection with consistent results for data recovery. Generally, a SLA, either formal or informal, exists between the backup operations team and the business units it serves. In order to have a fighting chance to deliver against a realistic service level expectation, the operations team must have extensive information available about the status within the data protection environment. At any given point in time, the backup operations team needs to be able to answer the fundamental question: 'How well are we protected?' This is a powerful question, full of many possible subjective interpretations. The goal is to take the subjectivity out of the process and establish some agreed upon metrics that can be compiled to determine the answer to this question.

One standard metric for determining the level of protection is to measure the success rate of backup jobs. The method of calculation used to measure this varies from organization to organization, but is generally some derivative of the formula

$$\text{all jobs success rate} = \frac{\text{number of backup jobs successfully completed}}{\text{number of backup jobs attempted}}.$$

At this fundamental level, this metric provides a raw measure of success or reliability within the backup environment. This metric is often challenged, however, because it can produce a number that is misleading, given that it takes into account jobs that have failed, but have been restarted and ultimately succeeded. All jobs success rate may more appropriately be used to measure the overall efficiency of a backup operation, rather than a net level of protection. For example, if an organization consistently operates at a 50 % all jobs success rate, it clearly experiences some problem(s) that result in half of the backup jobs failing.

Considering that this metric takes into account jobs that have been retried, perhaps multiple times, and may have ultimately completed successfully, it is possible that the percentage of data that is recoverable at the end of a backup window is actually much higher than what is indicated by the all jobs success rate. Although it is important to track the all jobs success rate because it can signal problems within the storage environment, servers, applications, SAN fabric, tape devices, backup application and so on, a more appropriate measure of the level of protection at any given time might be the *last jobs success rate*:

$$\text{last jobs success rate} = \frac{\text{number of last backup jobs successfully completed}}{\text{number of last backup jobs attempted}}.$$

This metric measures success based on the last attempt for each scheduled backup job within a period of time. In other words, last jobs success rate factors out jobs that initially failed, but through some measure, either an automated retry or manual intervention, the job ultimately completed. To illustrate the difference between these two metrics, consider the following example.

A backup job is scheduled to commence at 9:00 p.m. In the first three attempts, the job fails for a variety of reasons, but the fourth attempt completes successfully. By calculating the success of this backup job, we get two dramatically different rates:

$$\text{all jobs success rate} = \frac{\text{one job successfully completed}}{\text{four jobs attempted}} = 25\,\%,$$

$$\text{last jobs success rate} = \frac{\text{one job successfully completed}}{\text{one job attempted}} = 100\,\%.$$

In this scenario, both metrics are important to consider. First, a low all jobs success rate means that problems exist in the environment. This statistic alone does not give sufficient detail to begin to understand the nature of the problem, but it should provide incentive to begin some investigation. In order to do this effectively, the backup team needs access to additional tools which will be discussed later in this chapter. Until the problems are addressed and the all jobs success rate is raised to an acceptable level, the organization will be challenged to consistently deliver a high last jobs success rate, and the cost of protecting data is probably out of proportion with other areas of IT expenditures. The second metric, last jobs success rate, indicates that 100 % of the data is protected up to the time of the last backup. This is generally presented as a 'business facing' metric, as it gives assurance to the business that its data is protected.

The backup operations team also needs to be able to easily access data within the backup environment to use for tasks such as troubleshooting failed jobs and capacity planning for disk-based data protection or tape drives. In order to accurately do this, it is important that the backup operations team establishes real-time or near real-time visibility into the entire backup operation. Ideally, this visibility includes all master servers, media servers, clients, tape SAN infrastructure and tape drives/robotics. This dashboard view should also collect errors and warnings and provide a summary view of information with quick drill-down capability so that operators can quickly troubleshoot and resolve problems that exist (see Figure 5.2).

5.4.2 Alerting and Notification

To deliver a high service level, a backup team needs to be able to respond quickly to errors that occur in the operation. Failure to respond quickly

Figure 5.2 Backup operators console. Reproduced by permission of Symantec Corporation

to a critical error exposes the organization to risk and extends the window of vulnerability as the time lengthens since the last good backup copy was created. If a backup operator is expected to respond quickly to a problem occurring during backup, immediate notification of the error must occur. This can be accomplished by implementing sensors that monitor critical points in the operation and trigger some method of direct, proactive communication with the appropriate members of the team when a critical threshold is met. This notification can be sent via email, pager or Simple Network Management Protocol (SNMP) trap that is incorporated into a larger framework monitor with its own notification system. Once a notification is received, the backup operator has a chance to respond and remedy the underlying problem within the SLA window, provided that the environment is adequately tuned so that the number of error notifications does not overwhelm the backup team.

This technology could be used in the following way: A backup team is responsible for protecting 100 terabytes of data, housed within a centralized tiered storage environment consisting of NAS and SAN storage. Tier 1 storage consists of 20 terabytes of high-performance RAID 1 (redundant array of independent (inexpensive) disks 1) fibre channel disk arrays, used for storage of SAP data. Given the mission critical nature of this data, it receives the highest service level from the backup team. Therefore, hourly snapshots are taken, with off-host backup of Oracle logs. Nightly incremental backups are performed, again off-host to minimize the impact to the SAP system. Numerous sensors are implemented in this environment. If a snapshot fails once, the backup administrators are emailed. If a second consecutive snapshot fails, the backup administrators are emailed again, along with the group manager, who is charged with escalating and manually securing an acceptable form of backup without disruption to production. Sensors are also placed on all of the tape drives used to back up the SAP data, and administrators are notified when these drives fail or when a drive has been in use for 85 % of its MTBF, so that proactive maintenance can be performed on the drive. This team uses alerting and notification to focus its attention on the data with the highest priority. Alerts are also set for the backups scheduled for the other 80 terabytes, but the alerts are not based on as stringent thresholds as for the tier 1 data. For example, the tier 2 data may not flag a problem until the last attempt of a scheduled backup has failed, with no notification at all for failed snapshots, which are captured in a log and visible from a dashboard view, but not escalated aggressively.

5.4.3   Backup Reporting to Business Units

As SLAs are forged between IT and business units, pertaining to the backup and recovery of data owned by the business unit, business units require some level of assurance that the IT organization is upholding its commitment. Although IT metrics such as all jobs success rate and last jobs success rate are used as the gold standard for internal IT measurements of data protection efficiency and effectiveness, these metrics are not meaningful to a business unit without additional qualification. In fact, the all jobs success rate is entirely worthless to the business unit, as the business unit only cares whether or not its data is ultimately recoverable, and the measure of how many attempts it took is not relative to business (other than perhaps the alarming risk and high cost of constantly trying to overcome a high failure rate). At a first glance, however, the last jobs success rate would seem to provide a fair proxy for the level of protection, or in other words, the organization's ability to recover.

One might expect that with a last jobs success rate of 98 % or higher, a business unit is adequately protected, but in fact that statistic may be misleading when we examine the interests of the business unit.

The business unit does not have any inherent interest in how many backup jobs were run and how many were successful. It also does not necessarily have any knowledge or interest in how many terabytes were backed up or how long it took. Its interests are simple and nontechnical. It owns some applications, and if something bad happens to the data within an application, it needs to know that IT can reliably recover that data within an agreed upon period of time. So why does not the last jobs success rate fairly represent this? To understand this, we need to consider how applications are backed up. As an example, consider a content management application used to support prepress production in a printing and publishing company. This content management system consists of the following components: an Oracle server, two web servers, an application server, a middleware server and six NAS devices to store the actual content. In order to protect this application, the backup team schedules backup jobs to occur at midnight on each of the systems. To back up all of the components of these servers on varying server platforms, let us say that 100 backup jobs will be created with the intention that in aggregate they will protect the application. This number of jobs may need to be created to optimize performance and load balance the I/O across available resources. On this particular night, all of the backup jobs queue at midnight, but for some reason one of the jobs pertaining to Oracle database fails in all three of its scheduled attempts. The standard IT measurements of success indicate a 97 % all jobs success rate and a 99 % last jobs success rate. Unfortunately, the single backup that failed was a critical table within Oracle, and without a valid backup of that table, the entire database backup is worthless because it is impossible to recover the database to a consistent state without the critical table. And without a backup of the database, there is no possibility of recovering the application to the point in time of the backup, so all of the other backups are worthless as well. So success as measured by recoverability to the business unit for this application is a binary; either it is recoverable or it is not. In this example, the business unit measures success as 0 % and is exposed to up to 48 hour's worth of lost data changes and additions if the problem is not remedied and the backup reinitiated.

So how can a business unit measure recoverability in its own terms? The business unit needs to have IT metrics translated into terms that are relevant within the context of the business unit. The areas of primary concern within the business unit are recoverability of a business process, recovery point or risk exposure, recovery time and disaster recoverability. Let us examine each of these in greater detail.

Recoverability of a business process. This will consist of one or more critical applications that serve as the underpinning for a defined operation within the business unit. For example, the content management system described in the backup example may be only a part of a suite of tools used for workflow within the prepress and publishing operation. The workflow may have dependencies on other applications, such as CRM or a job routing system, and the net result of a failure in any of the applications is that eventually production comes to a halt in all areas (see Figure 5.3). Therefore, a business unit needs to be assured that the business process itself is protected, requiring representation of backup data at a business process level. This is a nontrivial task, but in such an operation where downtime has such far-reaching implications, it is worth the investment in a tool capable of mining the data and presenting it to the business unit in such a fashion.

Recovery point or risk exposure. The business unit needs to monitor its fallback point. How much data is exposed to loss if an array malfunctions or a corruption occurs? Depending upon the criticality of an application, a loss of a single hour's worth of production data can result in the loss of exorbitant amounts of money, and several hour's worth of lost data can potentially be catastrophic to the organization. Not all data carries the same impact, but it is important to regularly

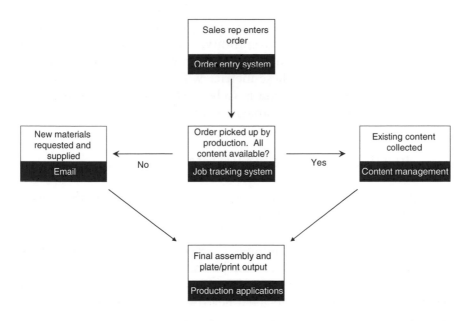

Figure 5.3   Interdependency of applications within prepress workflow. When one of the applications in the critical path becomes unavailable, the entire process can be held up

measure the risk associated with each application so that timely escalation can occur in the event that unacceptable data loss exposure emerges on any application.

Recovery time. If data loss or corruption occurs, how long will the business unit have to wait until the application and ultimately the business process can be back on line? This is a function of several key variables. Is the backup data on disk? Is it on tape? If on tape, is it currently resident within a library environment? If so, how many GB are necessary to recover the application and how many GB per hour can be recovered on average by the backup operations team? Also, what is the average wait time before a queued job can reach the top of the priority list and receive tape drive and media server resources to begin a recovery job? All of these factors can be combined to present an estimated recovery time capability to a business unit, which can properly set the expectations and the business unit will have assurance (or not) that the recovery time SLA can be achieved.

Disaster recoverability. Another function of data protection is the ability to recover data in the event of a site disaster that destroys primary data as well as its associated backup copies. If replication is used, how current is the off-site replica copy? If off-site tape rotation is used to establish disaster recoverability, what is the lag time between the time a backup job commences and the time the tapes are picked up by the delivery service? These metrics will provide an indication of exposure window, where data may be backed up for local recovery purposes, but remains vulnerable to a site disaster.

5.5   BUSINESS UNIT CHARGEBACK

As data protection metrics become translated into terms that relate directly to individual business units, and measure and monitor the effectiveness of the protection of their data, this information can also be used to establish accountability for utilization of a shared IT resource. This accountability can be taken as far as charging the business units for the portion of IT resources that it consumes. This level of accountability can bring a healthy balance to the negotiation over SLAs between IT and the business unit, as IT can deliver higher levels of service if the business unit is willing to fund the investment which would enable it.

Once a business unit begins to get charged for the services provided by IT and is given reporting level visibility into the backup operation, the organization may allow it to shop the open market for competitive services if the business unit feels that it is being overcharged or under-served. This trend has motivated many internal IT organizations to establish a more clear understanding of its cost structure and to aggressively drive costs out of its operation.

5.5.1   Backup Service Providers

IT outsourcers may offer data protection as a service along with other offerings. As this happens, it will compete with internal IT organizations and will have to commit to delivering a superior service level or lower cost in order to win and retain this business. In doing this, it must first understand its cost structure, including any component necessary to delivering data protection to its customers. These costs will include some portion of facilities, hardware, software, maintenance, liability as well as consumables such as tape media that are allocated to a particular client. On top of all of its costs, it needs to build a profit margin into the price extended to the customer. All of these factors do not translate neatly into a pricing model that the customer will expect, but yet it must be done in a way that the price can be competitive against internal IT as well as other outsource service providers. The commitment made by the outsourcer to the customer is generally in the form of a cost per gigabyte and is based on

- the total number of gigabytes protected;
- the retention time for backups;
- the customer's RPO and RTO parameters.

In order to displace an internal IT organization in the data protection operation, the outsourcer generally needs to demonstrate superior service and significantly lower costs. To deliver to this level in a sustained way over a long period of time, the outsourcer must maintain an efficient infrastructure and use state-of-the-art technologies so that it can provide the same level of service with fewer administrators per terabyte than the internal IT operation. It also must provide a greater degree of transparency into its operation so that the business customer can remain in a constant state of assurance that the data is protected to levels specified in the SLA.

To harvest maximum efficiency from its operation, the IT outsourcer commonly shares infrastructure components across multiple customers. While helping to minimize its own infrastructure, this practice makes it difficult to allocate costs fairly to specific customers using a portion of a shared resource. This necessitates system auditing and reporting tools that maintain granular visibility and allocate resources according to flexible definitions established by the outsourcer. That visibility needs to be given to the customers and ideally is also used as a direct input into an accounting system which calculates the costs of serving each customer based on the portion of the infrastructure and operation that it consumes.

5.6   CONCLUSION

Rapid growth in data, against a backdrop of flat IT budgets, has placed increasing strain on backup administrators who are expected to continue to deliver consistent levels of service. This considerable challenge requires careful analysis and the deployment of strategic management and reporting technologies designed to increase the administrator's level of control over the backup environment. These technologies can be further leveraged to provide transparency into the data protection operation for business units and to proportionately allocate data protection costs back to appropriate P&L centers. In the next chapter, we will examine security, which is another factor that plays an important role in data protection, and data management in general.

Chapter 6

Security

6.1 INTRODUCTION

The term 'data protection' was originally conceived to refer to backup and restore operations. Backup 'protects' the data from loss or corruption by making a secondary copy of the actual bits available for restore. Furthermore, this backup copy is separate and different from a mirror copy, which is not useful in the case of data corruption. Today, the term is increasingly encompassing the actual protection of the information contained within the data from unlawful access and theft, in recent months several highly publicized incidents regarding loss of nonencrypted data have created an unusual level of interest in encryption. In many of these cases, backup tapes containing data were lost during their transfer by truck from one location to another or stolen. This not only triggered a wave of legislative actions but also prompted data protection vendors and some security startup companies to look into providing solutions to address this problem.

The problem of securing data at rest is only one of multiple security aspects of data protection. Access control and audit logging have always been a fundamental component of operating systems and to some lesser extent data protection. Furthermore, the new architectures of IT organizations with firewall structures in place have put an additional burden of operability on data protection application design, development and implementation. In the rest of this chapter we will briefly go over encryption and the role of encryption in data protection applications. We will also briefly touch upon security

Digital Data Integrity   David Little, Skip Farmer and Oussama El-Hilali

functionality in data protection applications such as role-based security and audit trails.

Another subject that we incorporated in this chapter is the phenomenon that has been increasingly plaguing software development which is security exploits that result from weaknesses or vulnerabilities in operating systems or applications. This phenomena created a whole entire cottage industry of hackers who are discovering these vulnerabilities and others who are taking advantage of them with malicious intent.

6.2   ENCRYPTION AND DATA PROTECTION

6.2.1   Encryption Overview

Encryption is the process of converting readable and recognizable data into unreadable data in order to conceal the information from everyone except those who have the key to decrypt it to its original state. The encryption process requires a 'key' to ensure that the process is consistent and the encrypted data can be decrypted. Early civilizations such as the Egyptians, Assyrians, Chinese, Greeks, Romans, Arabs and others have used encryption to protect trade secrets, government communications, military plans and other sensitive information.

Today, the need to encrypt data and conceal sensitive information is more prevalent and accessible through the use of computers that can provide sophisticated encryption/decryption algorithms. In its most basic mathematical form, a string of data is subjected to an arithmetic operation against a 'key' to produce an encrypted string. The encrypted string and the key hold the information which can be reversed to its original state by applying the reverse of the initial operation.

In other words, if we wanted to encrypt the word 'Hello', we would read the string character by character: H, e, l, l and o.

If we take the ASCII value of each of these characters (72, 101, 108, 108 and 111) and add a constant such as 7 to each one of these ASCII values, we obtain the following: 79, 108, 115, 115 and 118.

By converting these ASCII values to their character equivalent, we obtain the following string: Olssv.

In this case, the encrypted word is 'Olssv' and the encryption key is the number 7.

To decrypt the message, we would read the encrypted word 'Olssv' character by character: O, l, s, s and v. After obtaining the ASCII values of those characters, we subtract 7 from each number and when we convert the ASCII values to their character equivalent, we obtain the original string 'Hello'.

We can add a higher level of sophistication by using multiplication and division instead of addition and subtraction. A further level of sophistication would be to use scalar multiplication with invertible matrices. For example, a character string can be divided into three character substrings and each of the ASCII values of the substring's characters is multiplied by an invertible matrix. Converting the ASCII values for their character equivalent produces an illegible or encrypted string. They key is the invertible matrix. In order to decrypt the resulting string, we reverse the operation and use the inverse of the matrix which was used to encrypt as our key to decrypt.

Before we discuss how encryption is used in data protection, let us summarize this section by emphasizing that encryption is composed of two parts:

1. The encryption algorithm, which is known.
2. The key, which is kept secret.

In the next sections, we will see that the availability of sophisticated encryption algorithms alone is not sufficient to solve encryption problems in data protection. In fact, the most challenging problems with the use of encryption in data protection – one may argue – are related to the encryption keys and their management; if the key or keys are lost and the data cannot be decrypted, the resulting effect is exactly the opposite of what data protection is intended to do – protect the data from corruption and loss. Hence, the absence of data integrity.

6.2.2 Encryption and Key Management

One of the biggest challenges in encryption is key management. This includes the ability to provide adequate storage for the keys ensuring their safety, recoverability and the ability to have an audit mechanism that can keep track of the changes made to the keys. Some aspects of key management are included and are a part of the data protection application. However, other parts are dictated by the processes developed by the organization using encryption, which are in turn developed based on the needs of the business.

The aspect of distribution versus consolidation appears and reappears throughout the discussion of key management. For example, some algorithms require that the encryption key be made of multiple components. Each component resides physically separately from the others, let us say with multiple individuals. Bringing the components together constructs the key that decrypts the data. This makes sure that one single individual cannot decrypt the data and therefore provides a certain level of access control. However, by having multiple components we increase the risk of not being able to decrypt the data when we need to when one of the components is not present.

This same aspect appears in data protection in another form. Storing the key in the data application client provides the user of that client the flexibility in certain environments to restore their data and decrypt it. However, in a disaster recovery situation where the client data is lost, rebuilding all the clients will require manual intervention at each client – a very costly operation in large environments.

6.2.3   Encryption Use in Data Protection

In data protection, advanced encryption algorithms such as Data Encryption Standard commonly known as DES and Advanced Encryption Standard or AES using 128- and 256-bit encryption help address the following two issues:

1. Protect the data while it is moving from the client to the media server to its destination of tape or disk.
2. Encrypt data targeted to tape that is intended to be transported off-site.

Typically in a large organization, the concern over internal or external unauthorized access to critical machines and therefore sensitive data can cause the data protection administrators to look to encryption as a solution to this problem. Some data protection applications encrypt the data at the client automatically as it is being sent across the local area network (LAN) or wide area network (WAN). Other data protection applications provide the ability to encrypt the data at the client as an option with multiple choices for the strength of encryption and a choice for the encryption algorithm. This use of encryption can protect against the unauthorized access of the data while the data is moving across the network. It can also prevent the unauthorized access to this data when it has been backed up and residing on its secondary storage.

Unauthorized access does not necessarily have to be with malicious intent. For example, a data protection administrator may unintentionally restore data to the wrong machine. In other cases, a disgruntled employee seeking access to sensitive data may do so by accessing the target machine through the disguise of being the data protection application. If the data is encrypted and ends up in the possession of an unauthorized entity, the encryption will render it useless without the encryption key.

Organizations that have requirements to store the entire data or a subset of their data externally for disaster recovery purposes need to physically transport this data to the designated off-site location on a regular basis. This is typically done by a third-party service and the data is normally on tape media. If one of the tapes is misplaced or stolen and cannot be accounted for, the organization has to assume that the data on that tape has been compromised. This means that the organization has to take action to mitigate the risk of that data falling into the wrong hands. In some cases, the law mandates that certain steps be taken which may include the notification of individuals whose data is believed to have been compromised. This type of incident can be very expensive and can cause irreparable harm to the credibility of the organization.

When a data set consists of multiple tapes, not all of them need to be lost or stolen for the data to be considered compromised; one or more tapes could simply be misplaced or unaccounted for. This tape may contain a full image or part of an image. Furthermore, even if the tape falls into malicious hands, the data would have to be read using the application that wrote it and the application may need a number of parameters to fully decipher it. Still, nothing short of a reliable encryption solution can guarantee that the data is unreadable and therefore uncompromised. For this reason, many organizations are resorting to encryption to address this problem. In particular, they are looking for solutions that encrypt only data that is targeted to be removed from their data center and the secured confines of the organization.

Although the client encryption solution discussed above can provide this type of protection, many organizations feel that it is too processor costly and key management intensive. A solution targeting certain data with global key management functionality is felt to be more acceptable and appropriate. These specific requirements have sent many data protection application vendors, as well as a number of startup companies, rushing to develop

- standalone software solutions;
- integrated software solutions;
- supported third-party hardware solutions.

Some of the hardware solutions offer a box using a processor that can be placed in the path between the media server and the tape drive to capture the data and encrypt it. Because this kind of solution normally will not require the media server processor to do the encryption, it is considered to be more appealing in those environments where processor power at the media server is expensive or limited. However, the drawback of such a solution is that the decryption of the data depends on the box's availability. To reduce this risk, another box is introduced in the configuration and some sort of replication is added to create redundancy. In addition to these solutions, tape drive vendors and data application vendors are announcing tape drive based solutions that can work in conjunction with data protection applications to provide not only encryption but also a higher level of key management. Some vendors are even considering the host bus adapter (HBA) as a potential location for the encryption.

6.3   DATA PROTECTION APPLICATION SECURITY

The nature of data protection is such that it has access to all or the most important data in the enterprise. This access is required and provided so that the data can be protected in case the primary data is lost or unavailable. However, the task of moving large amounts of data on a regular basis has the potential to create situations where the confidentiality of the data may be compromised.

One way to reduce this risk is to limit access to the data to only those individuals who need to have access to perform specific data protection tasks. For a long time, some operating systems and file systems have provided their users with the ability to control access to the data through a combination of mechanisms that include role-based security, access control lists (ACLs) and audit trails.

With the advent of the Internet and the need to share applications and data with users who may not necessarily be known to the operating system through user accounts, more elaborate mechanisms have been developed for shared applications especially web applications. These mechanisms have been put in place to monitor and restrict unauthorized access to commonly shared applications. These mechanisms rely on structural notions such as firewalls and architectural components like authentication, authorization and access control. When working properly, these mechanisms combined with the security structures offered by the operating system are intended to protect the data from unauthorized access.

Most applications in general and data protection applications in particular today offer security mechanisms that are derived from and are supposed to work in conjunction with the general security concepts described above.

6.3.1   Terminology

6.3.1.1   Authentication

Although the concept of authentication can apply to users, systems and applications, in the application world it means the ability to securely and unambiguously establish the identity of a user, including the user's group and membership in any roles (see Section 6.2.6). When a user is authenticated, a certificate is issued. This certificate describes the identity of the user to the rest of the system. The certificate carries with it and exposes relevant information about the user. The major benefit of authentication is visible in enterprise environments where the data protection application is distributed over multiple systems and leverages other applications to perform its data protection tasks as a part of one integrated system.

6.3.1.2   Authorization

Authorization for the purpose of this discussion is defined as ensuring that an authenticated user can perform all those, and only those, operations within a product which the user is allowed to perform according to the assigned permissions assigned by the security administrator of that product. Authorization to perform operations is granted to a user by assigning that user, or any group of which the user is a member, certain privileges.

Privileges are permissions to perform certain actions or tasks within the application. They are not associated with specific objects. For example, a device management privilege would allow a user to access device control parts of the data protection application, without implying permissions to perform specific operations on specific devices. Privileges can be assigned to any security principal, including individual users as well as groups. Privileges supported by a product are the domain of that product itself and do not need to be shared across products.

6.3.1.3   Access control

Having the privilege to perform certain operations does not in itself allow a user to perform actions on individual objects. It is only one of the necessary conditions. After an authorization check confirms the proper privileges to attempt an operation, access controls determine whether the implied actions on individual objects are allowed.

Where object-level access control is required, permissions of individual users or groups to operate on specific objects are described in ACLs associated with those objects. ACLs can grant or revoke permissions to perform specific actions on an object. Possible permissions vary by object type and are thus inherently product specific. Like privileges, they do not need to be shared across products.

Standards for managing security rights like Lightweight Directory Access Protocol (LDAP) and Active Directory are becoming the de facto standards for managing email and other Internet applications. These standards help with the authentication of the user. Once the user has been authenticated, the application can use additional information about the user such as groups they belong to and privileges they may have in other systems and then determine to either grant them the same level of access or apply further restrictions.

6.3.2   Role-Based Security

In a system using role-based security, the security administrator has a range of predefined roles that can be used to group certain functions and privileges associated with those functions to perform specific operations and access to designated data. In a data protection application, the list of roles may contain roles such as backup administrator, restore administrator and operator. There may also be a configurable role that the user can use to customize a role that is based on selective privileges from the other roles.

The system may also have a list of privileges. These privileges are associated with functions or modules of the data protection application. For example, there may be a privilege called mount media and another one called dismount media that will allow the roles that contain the privilege to mount and dismount media. These privileges may be associated with one or more roles. In most systems, the application is aware of and in some cases has the ability to reconcile its user list with that of the operating system in which it is running and that of the network. Also see Section 10.8.

6.3.3   Audit Trails

If the data protection application has a role-based system, then it can also provide an audit trail function. The main purpose of this function is to log all or most of the major activities performed by the system and identify the user who executed those functions. A secondary purpose is to identify who performed what type of maintenance on the system. The latter if it exists would be rare and would be used by technical support organizations to identify changes in hardware configurations when trying to analyse a support case.

New compliance laws such as Sarbanes-Oxley have created a renewed interest and demand in this type of functionality. Users hope that during audits they can identify to the auditors the steps taken with each major operation related to data protection.

In a typical audit trail, a record is created each time one of the identified functions for audit trails is executed. The record contains the user name, a time and date stamp, and the function executed as well as the result of the executed function. The audit trail function in the application should normally provide the user with the ability to search for the actions of a specific user or the user names associated with the execution of an operation such as running a backup or a restore.

6.3.4   Firewalls

Firewalls are a permanent part of the IT structure in today's enterprise data center. They consist of software, hardware or a combination of both. The primary purpose is to prevent unauthorized access to the enterprise's networks – inbound access. Often they are used to prevent unauthorized access from the network to the outside world – outbound access. A specified security criterion filters both inbound and outbound messages.

Most data protection applications were developed to operate within the secured confines of the enterprise's data center that is inside the firewall. But today's changing nature of data protection has pushed data protection application vendors to accommodate for firewall support. Initially, these applications required a large number of random ports to be opened for the application to perform its functionality to pass through the firewall. However, these limitations are slowly being addressed and new functionality is being introduced in data protection applications with less demanding requirements such as one single port to address the application's communication needs.

6.4   SECURITY VULNERABILITIES IN DATA PROTECTION APPLICATIONS

The advent of the Internet created new opportunities for users to share data, processing power and leverage computers in way unprecedented before. However, it also gave the malicious users an opportunity to create various ways to destructively use this superior networking capability. The hackers began to take advantage of vulnerabilities in operating systems. Most of these operating systems were developed and introduced prior to the Internet. They had a minimal level of readiness an resistance to the types and level of attacks created through the growth of the personal computer market and the Internet accessibility.

Operating systems were in general the first target for hackers. While black hat hackers used exploits – a term referring to an attack on a computer system with a vulnerability, white hat hackers or also called ethical hackers have been continuously helping the development of security patches or fixes to address these vulnerabilities. When these vulnerabilities became more and more difficult to find in operating systems, the black hat hackers moved to a higher level of sophistication and up the stack to take advantage of vulnerabilities in applications. One type of applications that are very intriguing to hackers are data protection applications due to their natural access to critical data in an organization.

Like operating systems, the most reliable data protection applications may have vulnerabilities. Even when these applications are developed or re-written using secure coding practices and are subjected to penetration testing, they can still contain vulnerabilities. Penetration testing uses ethical hackers or white hats to proactively identify security weaknesses in systems and address them before black hat hackers take advantage of them. In recent years, the security industry has seen a phenomenal growth of both white and black hat hackers. In fact, some of the white hat hackers employed today by organizations that specialize in penetration testing used to be black hat hackers in the past. The growth of black hat hackers has significantly encouraged white hat amateurs or hobbyists to turn into professionals and charge for their work.

Although early instances of penetration testing began in the 1970s, most of the efforts were organized by governments and assumed a deliberate and organized attack by an enemy government or organization. These tests often assumed scenarios less challenging than what we see today. Today's penetration testing tests for commonly known vulnerabilities. The combination of penetration testing and the use of secure coding techniques are helping application vendors to proactively reduce vulnerabilities in their applications. However, many are addressed reactively after the product or the new version of the product is released. This situation requires that both the developer of the application and the user be continually vigilant.

6.4.1 Vulnerability Detection and Fix Process

Some application vendors have opted to proactively go after potential vulnerabilities in their products and hired third-party companies to continuously look for and identify any potential weaknesses that may exist in their applications. Once they identify the vulnerability, the development team can fix it and introduce it as a fix in maintenance release or a patch fix. This activity accompanies some announcement in the part of the product team indicating that a potential vulnerability has been identified and fixed and urging the users to download the fix and install it. This disclosure helps the company to reduce the risk of exposure to potential harm to a user's data.

However, not all vulnerabilities are identified in this manner. There are organizations today that hire white hat hackers to identify vulnerabilities in well-known applications. These individuals are often part-time free lancers who have knowledge of design and coding flaws in applications especially those applications developed a decade or two ago. Once a vulnerability is identified the organization provides the application vendor the information and asks the application vendor to fix it in a short time frame before the organization goes public with it. In this case the vulnerability and the fix are announced at the same time and the vendor provides a hot fix that can be downloaded and installed.

In some cases the application vendor is not given any time to fix it and the vulnerability is announced to the public leaving the vendor very little or no time to react. This opens the door for what is called zero day exploit. In this scenario hackers who have seen the type of vulnerability announced are a lot faster than the application vendor and develop an exploit to take advantage of the vulnerability before the vendor has had a chance to fix it. In 2004, Symantec conducted a study that showed that although the number of vulnerabilities discovered in 2002 and 2003 was the same, the time between the announcement of the vulnerability and the existence of an exploit narrowed significantly.

6.4.2   Types of Vulnerabilities

There are various types of vulnerabilities. Often the vulnerabilities in applications are a result of poor design or coding. Here are some most frequently cited types of application vulnerabilities:

6.4.2.1. Buffer overflows: Applications use buffers to store information especially when interfacing with the user or other applications. If the amount of data sent to a specific buffer is bigger than what that buffer can hold it overflows to adjacent buffers. This can modify the original contents of the buffer and corrupt it. Malicious hackers can take advantage of this condition and position instructions to land in the overflow area and execute instructions defined by the hackers.

6.4.2.2. Integer overflows: Applications use integer arithmetic to accomplish many tasks like counting. Integer-type variables are used instead of floating-point-type variables to save memory and speed up the processing. The amount of memory used by an integer is less than a floating-point. However, arithmetic operations may produce results that are beyond the size allocated to store that value, causing an overflow. Normally, an integer overflow should not create a weakness but in certain conditions, like we have seen above with buffer overflows, this overflow can be maliciously manipulated to cause harm.

6.4.2.3. Raceconditions: Race conditions are another type of potential weakness that can be exploited by hackers. These conditions are created when commands to read and write data are received at the same time and the computer attempts to execute them without regard to priority or sequence. Hackers can take advantage of these types of vulnerabilities and insert specific instructions they wish that machine to execute.

6.4.2.4. Format string errors: Applications use string manipulation functions to format input and output data. Hackers take advantage of these types of operations when they are done right to insert instructions into the format string to access areas in memory to maliciously misuse this data.

Most of the application vulnerabilities found today stem from vulnerabilities that exist in programming languages or poor programming and design or a combination of all. This is why many application vendor have gone back to review their application code and replace the code that is vulnerable. There are new secure coding techniques that programmers are adopting that allow software engineers to write a code that is free of the vulnerabilities known today.

6.5 CONCLUSION

In this chapter, we have seen how the concept of data protection is expanding to incorporate security. We have examined the recent events that are making encryption of secondary data a must-have for many enterprises. We also briefly looked at elements of security present in the enterprise and how they are interacting with data protection applications. In this overview, we provided definitions for common terms used in security and described role-based security, audit trails and firewalls and briefly introduced the reader to the area of security vulnerabilities.

Chapter 7

New Features in Data Protection

7.1 INTRODUCTION

Backup administrators are constantly looking for ways to optimize the process of data backup and recovery. Unfortunately, advanced methods require time and resource investment that may not be easily justifiable. However, two significant factors that contribute to the optimization and efficiency of backups and restores are

1. The maturity of the data protection field which allowed application developers to implement advanced algorithms that leverage data that has already been backed up and therefore reduce the amount of data redundancy.
2. The continuous decrease in disk prices which has opened up some opportunities to implement features that leverage the disk's random access capabilities to shorten the backup window and provide for fast restores.

One method that requires very little change from the traditional backup routines and can potentially render significant advantages over traditional backups is synthetic backups. In this chapter, we will look at the concept of synthetic backups, its evolution and how powerful it can be when it is leveraging disk for a part or the entire

---

Digital Data Integrity   David Little, Skip Farmer and Oussama El-Hilali

backup and restore operation. We will also examine various disk-based features and the efficiency they bring to data protection.

7.2   SYNTHETIC BACKUPS

The concept of synthetic backups is based on the theory that a full backup image and a set of incremental backup images could be synthesized to produce the equivalent of a full backup image exclusively by applying the incremental backup images on the full backup image. The by-product is a new full backup image that is called a synthetic backup because it was synthesized and not obtained by running a full backup on the original data. The synthetic backup is expected to behave exactly like a full backup.

This concept can be used to build a full synthetic backup image from a full backup and one or more incremental backups or to consolidate a set of incremental backups into one synthetic cumulative incremental image. Depending on the nature of the problem the administrator is trying to solve and depending on the flexibility of the application used for synthetic backups, the backup administrator can choose one or the other or both to optimize the process.

For example, an environment performing weekly full backups on Sundays and daily incremental backups the rest of the week can, by adopting a synthetic backup approach, skip the full backup on Sunday and instead build a synthetic full backup without interrogating the backup client servers and without taxing the network. Additionally, the backup administrator may decide to synthesize the daily incremental backups into a single cumulative incremental on Saturday to expedite the process of building the synthetic full backup on Sunday.

Figure 7.1 shows an example of how a full backup (A) is used along with incremental backups done Monday through Saturday to create a synthetic full backup (B), and consequently use (B) and the incremental backups of the following week to produce another synthetic full backup (C) and so on and so forth.

Furthermore, the administrator in this environment may choose to optimize the use of tape drives by building the synthetic backups outside of the backup window when normally no backups are allowed because of the performance tax they may cause to the client servers and the network bandwidth usage.

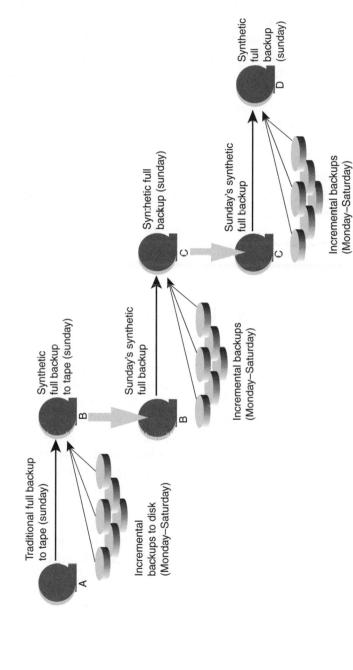

Figure 7.1 Synthetic backups

7.3   EVOLUTION OF SYNTHETIC BACKUPS

The continuous refinement of disk technology and its affordability allows disk to become an integral part of data protection. One of the areas that has benefited from the declining prices of disk is synthetic backup. Initially conceived to synthesize data residing on tape, the concept has evolved to take advantage of the direct access nature of the disk.

To synthesize data stored on tape, the synthetic engine must navigate the tape or tapes to collect the components of the synthetic image it is building. However, when the data is residing on disk, the engine now need not copy the data but can instead build pointers to that data saving a great deal of time.

In Chapter 8, we will talk about new and advanced synthetic engines synthesizing data residing purely on disk. In this chapter, we will discuss the more common use of synthetic engines that deal with data residing on tape or partially tape and partially disk.

7.4   BENEFITS OF SYNTHETIC BACKUPS

One of the most significant benefits of synthetics is the ability to optimize restores especially when a full recovery is needed. Normally, a restore operation consists of restoring from the most recent full and then applying the incremental backups after the full was taken in a chronological order. Would it not be nice to have a full available based on the last day's backup without having to run a full everyday? By building a synthetic full at the end of each incremental, the users can have at their disposal a full synthetic everyday or as often as they run synthetics for a quick restore that would not require the aforementioned process.

Because the processing required for building a synthetic full or synthetic incremental uses data already obtained from interrogating the client, there is no need to further interrogate the client when building the new synthetic backup. This means that the data protection process can proceed even when the client source is down or up but not accessible for backup operations.

The synthetic full is built by the media server and the data input in this process is previously backed up data not residing in the client; therefore, there is virtually no use of network bandwidth associated with interrogating the source of the data. Inaccessibility of the client does not prevent the synthetic engine from building a synthetic full. This ability can be very important when applied to remote office backups, for example. In many cases, remote offices are connected to data centers, but through very low bandwidth connections. The system administrators of such environments resort to putting a tape drive in the remote office due to the unreliability of the connection or simply the inability of the system to back up the amount of data needed to be backed up from the remote office or both. In addition to the cost of the tape drive, the management of the tape media in the remote office creates an additional cost to the system administrator and can add to the overall risk to the ability to restore and recover in a disaster recovery situation.

Depending on the nature of the data, synthetic backups quite often can help in these situations. After an initial full backup, only incremental changes would be sent to the data center and frequent full backup sets would be synthesized based on the initial full and subsequent synthetic full backups and the incremental changes. In these cases, the fact that no full backups are being conducted over the low bandwidth connection may just make this solution adoptable. However, before adopting such solutions one must examine the nature of the synthetic engine and the level of granularity for synthesis.

Another benefit is that normally backup operations run at night or after hours. This is done to restrict the backup operations from competing for network bandwidth and processor cycles of the machine being backed up. This way the data center optimizes the uses of its resources by dividing the processing into operations during the day and backup and maintenance during off-hours. In this type of situation, synthetic backups can help the optimization theme by building the synthetic full backups during the day and utilizing the tape drives during a time when the majority of tape drives would not be normally used.

In environments where a full is desirable to have everyday but the backup window is too short to accommodate a continuously growing set of data, synthetic backups may provide a suitable solution by which incremental backups are run on a daily basis during the available backup window but a full synthetic is developed during the day providing the environment with a desired daily full without increasing the backup window of operations.

7.5   BUILDING A SYNTHETIC BACKUP

In an environment where the synthetic engine deals with data entirely residing on disk, the engine may not need to move data within the secondary storage to build the synthetic backup image and may use pointer manipulation to build the new synthetic backup image. However, in environments where the synthetic engine has to deal with data stored partially on tape and partially on disk, manipulating pointers to build the new synthetic is not an option especially if the target is intended to reside on tape, and therefore the engine has to rely on a more I/O intensive process described in this chapter.

In the simplest case of building a synthetic full from a full backup and one incremental backup, the synthetic engine would parse the full to determine what the image looks like, then it parses the incremental backup to identify the files that have changed since the full backup was made so that they can be added to the synthetic full. This operation allows the synthetic engine to build a map with the location of each file so that at the time of building the synthetic full it can go directly to the location of the desired file and copy it to the target. Once the files have been copied to the new location, the engine updates the catalogue to reflect the new composition of the synthetic full.

When parsing the incremental backup image, intelligent synthetic engines have the ability to determine which files have been deleted since the full backup was run. This way those files are not copied to the target. Engines that do not have this capability can cause the synthetic backup to continuously grow exceeding the size of the primary data.

When adjusting the catalogue to reflect the changes made by the synthetic engine, it is important to take into consideration the order in which the catalogue was updated. In some implementations, the alphabetic order is respected so that search operations are easy. After the synthetic engine manipulations, the catalogue ends up looking like it would have looked after a regular full backup. However, this may not be possible to do without a significant amount of data movement, and therefore in some cases the order of the catalogue is reversed chronological order which enhances the performance of the engine but may require a postprocess that can rearrange the catalogue's data.

7.6 TECHNICAL CONSIDERATIONS AND LIMITATIONS

7.6.1 File-Based Versus Block-Based Synthetics

The concept of synthetics is often applied at the file level. This means that the synthetic engine uses the file as the smallest unit of work. It monitors the change attributes to a file to determine if a change has happened or not. In these circumstances, even the smallest change in file can trigger that file to be included in the incremental set and therefore be used to overwrite the previous version of that file when building the synthetic full. The implementation at the file level makes synthetics impractical for many situations where the file is very large like in the case of databases.

However, new implementations of the synthetic backup theory realize that quite often the changes in a file are limited to a few blocks not the entire file and therefore the synthetic engine uses the block as a unit of work and not the file. Because a block of data is normally smaller than a file, the data associated with the incremental backup will contain less data than an incremental of changed files.

7.6.2 File Types and File Change Frequency

Not all environments and types of data can benefit from synthetic backups. There are three parameters than can affect the efficiency of the synthetic backups:

- the size of the files;
- the number of files;
- rate of change of the data.

The next two examples are two extreme examples where the structure of the data (number of files) and to some extent the rate of change are factors that influence the success rate and benefit of the synthetic backups. We will see how the first environment can greatly benefit from synthetic backups, whereas the second environment is likely to experience no benefits at all.

Imagine a driver's license agency where the data consists of a large number of small files. Each file is a record representing the driver's license record of an individual. The files change only when an

individual renews their license. So basically the number of files changing on a daily basis is relatively small in relation to the total number of
files. A backup schedule that involves running daily incremental backups followed by building a synthetic image using the daily incremental
backups can provide the administrator with a synthetic full. This
synthetic full provides the administrator with all restorability of a
full backup on a daily basis and therefore tremendously decreasing
the restore time and eliminating the need to run full backups.

Now imagine the same agency where the information about an
entire community (group of driver's license holders) is all stored in
one large file. The renewal of a single individual's record can cause the
file to be marked as changed and therefore backed up during the
incremental backup. Basically, the incremental is the same size of a
full. In this case, a synthetic backup is not likely to bring additional
benefits.

7.6.3   Media Considerations

When creating a synthetic full, the synthetic engine deals with three
entities:

- the base full backup (either original full or synthetic);
- the incremental backup;
- the new synthetic full.

The synthetic engine will require access to all the images during the
process of building the new synthetic and if these images are on tape,
the engine will require either a large amount of memory or substantial
amount of tape drives to avoid mounting and dismounting tapes.

For this reason, some data protection applications either rely on disk
to complete the synthetic backup or provide an option to leverage disk
in the process of building a synthetic full.

7.7   DISK-BASED SOLUTIONS

As disk becomes cheaper, more data protection applications are beginning to take advantage of the random access capabilities disk offers in
order to optimize the backup and restore operations. However, these
solutions often require significant investments in a new infrastructure.

Organizations that have recently invested in a tape-based infrastructure may be more interested in integrating disk within their environments rather than replacing their existing environments with the new ones.

For this purpose, many data protection application vendors have built solutions that require very little or no new investment by the user to introduce disk into a traditional tape-based environment. In this section, we will go over some of these solutions emphasizing their benefits and underlining their limitations.

7.8   DISK TO DISK

For a long time, the data protection pundits talked about fast backups and promoted ideas and methods that concentrated on shrinking the backup window. However, more recently, the backup window became, in the eyes of many users and analysts, secondary to the speed and reliability of data restores. The reason is that in a disaster recovery scenario the downtime can be extremely costly to a company and in many cases if the data is not restored on a timely basis, the business may never recover.

Interestingly enough, this emphasis on recoverability seemed to parallel the emergence of disk as an affordable medium for data protection. As users began to consider disk media for backup, they began to discover that restores from disk were much easier and faster in certain situations. Disk offers random access which allows faster access to the desired blocks of data during restore operations especially when the restore is limited to one or a few select files.

Take for example the situation where a backup administrator in a large enterprise receives a request to restore a file that was inadvertently deleted by the user. Regardless of the media used for backup, some steps taken by the backup administrator remain the same. For example, running the restore application and locating the file are not very much affected by the type of medium used. However, the steps that ensue can be significantly different and provide faster recoverability for disk. In the case of tape, the application identifies which tape has the file and requests the tape to be mounted. Once mounted, the tape is spun to the location where the file is stored, the blocks are read and the tape is dismounted.

When the data is stored on disk, a simple call may be enough to get the operating system to locate a file on the disk and read it in seconds.

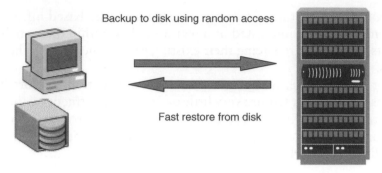

Backup to disk using random access

Fast restore from disk

Figure 7.2   Backup to disk

This is a significant time reduction when compared with the tape process which may take minutes and in some cases may have to rely on the availability of a tape drive for the restore operation. Now imagine a large-size company where the backup administrator receives hundreds of these kinds of requests per day. The difference between tape and disk can significantly change the nature of the service level agreement (SLA) the backup administrator offers.

The realization of the powerful benefits of disks has caused many to innovate in this area, producing more and more efficient and refined technologies. One such technology is MAID, which stand for massive array of idle disks. This technology, which utilizes low-cost serial advanced technology attachment (SATA) disks, provides an effective way to leverage disk for data protection without the potential high cost of maintenance. Because of its ability to use a large number of disk drives in which only those drives that are actively used are spinning, MAID is able to reduce the power consumption and prolong the life of the disk drives. A MAID can have hundreds and even thousands of drives thus competing with the low- and well-established world of tape.

Figure 7.2 depicts a disk-based backup solution that backs up primary data from a server into an array providing fast backup and restores.

7.9   DISK STAGING

The benefit derived from the usage of disk is limited by the timing of the restore and the retention period of the backup data. Various users have shared with us that most of the restores happen during the early days of the retention period with a high number of their restores (80–90 %) happening during the first two weeks following the backup. As disk is more attractive for restorability than it is for long retention, there is a balance between how much of the early retention period needs to go on disk and how much should go on tape.

This is where disk staging comes in handy by offering a means to back up to disk so that the data can be easily accessible for a fast restore during the period of time when the restores are frequent, then move the data to tape when the restore demands on the data have declined.

The concept behind disk staging is conceived for and is mostly applicable to those environments where the users like to take advantage of the random access capabilities of the disk to obtain faster restores. However, implementing a complete disk-to-disk backup solution is costly or may not be desirable due to the need to take the data off-site. For these environments, a mixture of disk media and tape media for the backup data offers the ability to decrease the restore time without incurring the cost of a total move to disk.

7.9.1   Early Implementations

One of the earliest home-grown implementations of disk staging was to back up the data to disk, then through a script make a copy to tape that would be shipped off-site. Then maintain the original copy on disk and refresh that copy frequently deleting the old version and replacing it with the most current backup. Although this process makes the restore for the data residing on disk faster, it actually may delay restores that are coming from tape that contains backup data copied to it from disk using the script. When the file or files needed are no longer on disk and have to be restored from the tape, the user must apply a reverse script to the one that copied that data to tape in order to copy it back to disk where the backup application could recognize it and restore from it.

This additional step is caused by the fact that the application that wrote to the disk has no knowledge of the copy that has been made to tape using the script. Some data protection application vendors were quick to remedy this by adding the script logic and functionality into the application making it aware of its location on disk for the initial backup and providing functionality to copy it to tape based on the date or the size of the disk cache available.

Figure 7.3 shows an example of early two-step disk staging that requires a two-step restore.

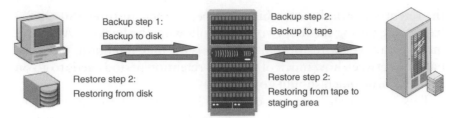

Figure 7.3   Two-step disk staging

7.9.2   Later Implementations

In addition to offering simple staging, data protection applications either have built or are planning to build advanced functionality to handle many of the issues surrounding disk staging. For example, users realized that it would be less time consuming if there was a function that allowed the backup to be written simultaneously to disk and tape. This way the tape backup can be sent off-site if necessary immediately upon the end of the backup, whereas a copy of the data remained available for restores. This functionality, also known as twinning, need not be limited to one copy on tape and one on disk. Multiple tape copies can be obtained allowing for one copy to remain in the premises and the others to be shipped off-site. The copy remaining on-site can be used by those users that may have a short retention period for their disk staging.

Another issue associated with the staging functionality is the management of the disk cache used for staging. This cache is often limited and requires that data written to it be recycled on a regular basis. One approach is to recycle the data based on the capacity of the cache. This means that once the cache is full, a process is launched to expire backup images on a chronological order basis, meaning keep deleting old backups until you have enough room to write the new one.

This method makes an optimized use of the cache, hence maximizing the restorability from disk. However, expiring old backups can be time consuming especially when the application is ready to write a new backup. First the application needs to figure out how much disk it needs for the backup it is about to write, then it needs to expire as many images as needed to free space for the new backup, all this happening during the backup window using time that could be otherwise used to do more backups.

An alternative approach is to recycle the data based on a retention policy. For example, the data protection application would allow the user to set up a policy to allow the user a number of days worth of backup images. At the end of each successful backup, the application expires the oldest backup image. Although this method does the expiration at the end of the backup window and not at the beginning of the backup, it too has its shortcomings. The disk cache is not used in an optimal fashion as the expiration happens at the end of the backup. There is always a certain amount of empty disk that could have been used to have one more image available for restore. Furthermore, this method is subject to inconsistencies in the amount of images left. If one of the backup images suddenly grows beyond its anticipated size, it will require the application to remove more than one image to make room for the next day backup. This in turn reduces the amount of data available in the cache for restores.

Advanced data protection applications use the concept of watermarks. This concept simply allows the user to identify a high watermark and a low watermark. The high watermark identifies the maximum amount of images that can fit in the disk cache before the expiration process would start to run and expire enough images to bring the contents of the cache down to the low watermark. This low watermark is also the indicator for the minimum amount of disk space that will always be reserved for the images to provide optimal restorability. Some functionality can be combined by an option to schedule the expiration process which then allows the user to expire images and bring the cache down to the low watermark at a time of their choosing outside of the backup window.

Figure 7.4 shows an integrated disk staging solution that allows the user to schedule the backup, staging and transfer to tape using the data protection application.

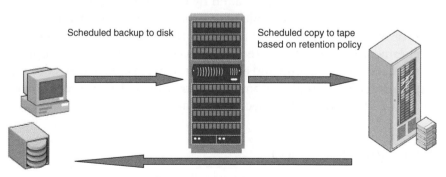

Figure 7.4   Disk staging

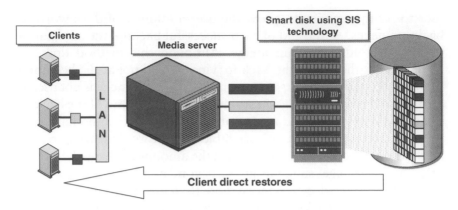

Figure 7.5   Commercial implementations of disk staging

7.9.3   Commercial Implementations

Some disk vendors have partnered with data protection application vendors to provide a bundled solution that offers the customer optimized disk usage with their preferred data protection application. These types of solutions are generally meant to provide a competitive throughput because in theory the application has been tuned to the parameters of the disk.

In some cases, the application vendor has implemented a special code to take advantage of additional data reduction techniques such as single instance store (SIS), discussed in Chapter 8, and compression features that the storage device offers. In this case, the advantage offered by the bundle is great because in addition to the benefit offered by the disk media, the actual size of storage required to store the backup may be smaller than the space taken by the original data.

Figure 7.5 shows an example where the data protection application vendor was able to work with a third-party disk vendor to provide an integrated optimized disk-based solution. The data protection application schedules the backup and executes it, whereas the disk vendor has implemented a smart algorithm in the array providing SIS technology to reduce the storage of the secondary data.

7.10   VIRTUAL TAPE

Another practical and popular disk-based backup solution is virtual tape or sometimes referred to as VTL for virtual tape library. Although most of these devices emulate tape libraries, some emulate tape drives and therefore we will refer to them in this chapter as virtual tape appliances. These appliances emulate tape drives and tape libraries by leveraging disk to store backup images. They are called virtual tape drives or libraries because, from the point of view of the data protection application, they are expected to behave like a tape drive or tape library.

These hardware appliances normally consist of a computer often running a Linux application that emulates tape drives and a tape library. One of the main functions of the application is to write to the redundant array of independent (inexpensive) disks (RAID)-based array of disk drives that is often bundled with the appliance but sometimes can be purchased separately. In most cases, the user can increase the storage capacity by adding additional storage. Another function that is offered by the application in these hardware appliances is the ability to configure the storage into virtual tape cartridges and drives by specifying the capacity of the cartridge and the number of tape drives.

7.10.1 Advantages of Virtual Tape

Most backup applications use optimized ways to write and read data to and from tape drives. However, when a read or write operation fails, it is difficult for these applications to identify where the failure happened. Often the user is forced to investigate the various stages the data has to go through before it is written on tape. There are a number of culprits that can cause write and read errors, including the tape drive and the tape.

One advantage virtual tape offers is the fact that the data is written to disk which is normally in the form of RAID storage. Tape drive and media hardware related errors simply do not occur. This can lead to decreased maintenance and management costs especially in large environments where hundreds of tape drives are deployed and thousands of tapes are used.

Because the tape drives are virtual and the medium is disk in virtual tape appliances, the time usually spent on mounting and dismounting real tape can be eliminated, resulting in time savings and therefore shortening the backup window. This advantage is noticeable when restoring one or multiple files that normally would be located in multiple tapes. However, this advantage is less visible and may actually be a disadvantage in some cases where large amounts of data need to be restored like the case in disaster recovery.

Another advantage that comes with the virtualization in these appliances is the ability to define more virtual tape drives than the environment would actually need or use in physical tape drives and allocate the extra drives for restores. This type of arrangement provides for faster restore operations at no additional cost.

In general, these appliances are widely supported by the major backup application vendors. They are easy to deploy and provide flexibility with configuring the number of virtual drives and the capacity of the virtual cartridge.

7.10.2   Technical Considerations and Limitations

Perhaps, one of the most frequently asked question when considering virtual tape is performance. Most of the virtual tape appliances claim to provide 150 to even 200 MB/s throughput which is significantly higher than what tape drives can provide today. However, tape drives have been around for a long time and tape drive manufacturers have had an opportunity to refine various aspects of writing and reading to tape, including performance. One of the features that some tape drives offer is compression on top of speeds that can exceed 50 MB/s. In some environments where large amount of data is being backed up, these tape drives using streaming may offer an advantage over virtual tape.

Another consideration comes with sites that have processes requiring that data be stored in a remote site. These shops have a need to copy the data from the virtual tape storage to tape media. The backup application that moved the data from its primary location to the virtual media storage may not provide the function of moving data from virtual tape to physical tape. This adds an additional step to the process and therefore makes it a costly operation for these environments.

7.11   DISK-BASED DATA PROTECTION IMPLEMENTATION ISSUES

Although disk provides numerous advantages over tape, the usage of disk in data protection has its issues and limitations. Tape has had several years of exposure in this area and data protection application vendors have had the opportunity to optimize the read and write operations and take advantage of tape drive compression and other features offered by the tape drive vendors. As disk has not had that much exposure, the ability to find a mature data protection application developed from the beginning to leverage disk can be a challenging task.

The maintenance required for disk is obviously different from that of tape. Some tape vendors offer warranties of up to 30 years. Disk vendors have not generally been able to match that kind of warranty. Also once the data has been written to tape, the tape can be removed from the tape drive and stored. However, in the case of the disk, the situation is different. Disks require power to continue to spin. Powering them down increases the risk of their unavailability when they are needed for a restore.

Another factor that deters users from converting their environment to total disk is the availability of processes and third-party vendors who offer disk vaulting services. In the case of disk-based backup, there are no well-known methods of vaulting your data on disk and there are no known large vendors that offer this kind of service at this time.

7.12   CONCLUSION

In this chapter, we have seen how disk is increasingly leveraged in the data protection process by using advanced algorithms that are intended to optimize the backup and recovery process. We discussed synthetic backup technology and its main proposition values. We also discussed how disk is being leveraged for disk staging operations to provide faster restores and disk-to-disk backup in general. The third aspect of disk-based technology we discussed was virtual tape and how it is becoming a popular option for enterprise data centers that are using tape and choosing VTLs to migrate to disk.

It is important to note here that although these features can bring significant cost savings and improve the quality of the SLAs of an IT organization, they are not equally effective in all environments and bring with them a certain level of inevitable complexity. Organizations looking at taking advantage of these features must first assess the problem they are trying to solve and then choose the feature or technique that can help them best accomplish their goals. This is especially important when considering the next generation of data protection technologies that are purely based on disk that we will discuss in the next chapter.

Chapter 8

Disk-Based Protection Technologies

8.1 INTRODUCTION

In Chapter 7, we saw how data protection administrators are looking for new ways to optimize the use of a continuously shrinking backup window in order to provide cheaper and more manageable data protection. However, the changes in technology and the rapid adoption of affordable serial advanced technology attachment (SATA) drives in the backup and restore activities are providing opportunities to data protection vendors to explore new methods purely conceived with disk in mind. They are also providing opportunities to information technology (IT) organizations that are willing to trade their present predominantly tape-based environments for the new ones or at least significantly incorporate disk in the backup and restore process.

In this chapter, we will evaluate how the synthetic backup method can be further enhanced when designed to operate and synthesize data in pure disk environments. We will also look at how the rapid growth of data and the change of service level agreements (SLAs) are causing many backup administrators to look at continuous data protection (CDP) as a means to cope with the ever-growing requirements of restorability and compliance and still maintain reasonable SLAs. Finally, in spite of the affordability of disk media, the large amounts of data and point in time objective (PTO) and recovery time objective (RTO) are requiring data protection application vendors to consider

Digital Data Integrity   David Little, Skip Farmer and Oussama El-Hilali

methods that will store less data. One such method is single instance store (SIS).

8.2   DISK SYNTHETIC BACKUP

We discussed synthetic backups at length in the previous chapter. In the following section, we will cover a form of synthetic backups that is done purely on disk. This type of synthetic backup is often referred to as 'virtual synthetic backup'. Because this form of synthetic backup runs only on disk, it provides certain additional benefits over regular synthetic backup as described in the previous chapter. It is also noteworthy to mention that the disk synthetic engines will vary slightly over the engine described in Chapter 7, which is meant to accommodate both tape and disk.

The nature of tape imposes additional requirements on a synthetic backup engine that affect both performance and storage. When tape is involved in the process of building a synthetic backup, the synthetic engine needs to take extra steps towards the creation of the new synthetic backup image. For example, when the synthetic backup is created on tape, the original data from the full backup and the incremental backup has to be read and written to tape. Reading data from tape requires mounting the tape or tapes if the image is distributed over several tapes. Then, because tape is a sequential medium, the data has to be written in a certain order to avoid lengthy restores.

However, disk synthetic engines that are intended to get their input from disk and send their output to disk can take advantage of the random access nature of disk. Therefore, when building the synthetic image, the engine need not create an additional copy of the data but merely create a pointer that can refer to its disk location, without moving or copying it. This enhances performance and reduces the amount of storage required to store the backup data.

Take for example a situation where the synthetic engine is creating a synthetic full from a full backup stored on disk and an incremental backup stored on disk as well. The engine creates an index that points at the full backup and edits that index to incorporate the changes that have occurred in the incremental backup. Let us say that the full backup had files A, B, C and D. In the incremental, file C was changed. The incremental contains only file C. When the disk synthetic image is created, a new index is created based on the index of the full backup. This index will contain pointers to files A, B, C and D. After reading the incremental backup and realizing that C has changed, the index is modified to point at the location where the modified file C is residing. In this manner, the full backup is accessible and is preserved, and the disk synthetic full is available as well.

Disk-based synthetic backup engines are intended to provide better backup performance and faster restore capabilities. They are also intended to offer a more efficient backup method that takes less space by providing access to multiple versions of the same file without having to store all the versions. In this case, the synthetic engine stores only the blocks that have changed. However, this optimization does not come free of risk and limitations. The loss or corruption of a block may render all the backup copies unusable for restore, except when this risk is mitigated by creating snapshots or backing up to tape at the expense of the disk space savings. Also, it is generally assumed today that long-term retention needs will most often require copying data to tape.

8.3   ONLINE PROTECTION: CDP

The goal of traditional data protection, as described in the first chapter, is to make one or multiple copies of the primary data. Full copies are taken at longer intervals, and in between in smaller time intervals, the full copies are refreshed with what is called incremental backups. These capture the changes that have occurred since the full copy was taken. This approach has been adequate for protecting all types of data for many organizations but has provided some limitations that are now being challenged by new disk-based data protection technologies.

When we apply RPO and RTO metrics to traditional data protection, we find that the RPO is limited to the frequency of the incremental backups, which is normally between once every 24 h. Very rarely, organizations will do incremental backups more than once in a 24 h period simply because the process normally takes time and is usually done during an after-hours backup window. We also find that the RTO can extend from minutes to days depending on the data that is being retrieved and where that data is located. If the data sought for restore has just been backed up to tape, then the RTO is the time required to mount the right tape or tapes and copy the data off that tape or tapes. However, if the tapes have been moved off-site, then the RTO is the time needed to mount the tapes plus the transportation time from the off-site facility.

In addition to the delays caused by the identification of the tape media, transport, mount and dismount operations, the traditional nonoptimized backup and recovery methods on average have a relatively high rate of failure. Some analysts estimate that the rate of failure of tape backup is as high as 50 %. If you add this failure rate is associated with the tape drive technology, the average RTO for an enterprise data center is even worse.

With today's increasing dependence on electronic communication such as email, storage of critical imaging data in healthcare and minute-by-minute recording of business transactions in financial institutions, the RTOs and RPOs offered by the traditional backup are just not acceptable to these businesses. Faced with these hurdles, traditional data protection vendors, challenged by some startups, have gone looking for a more practical backup and recovery paradigm in which data is continuously backed up to secondary storage so that shorter RTOs and more granularity in RPOs can be obtained. This resulted in the production of a number of applications that claim to do CDP. In the next sections, we will examine a definition of CDP and a couple of ways used to develop CDP applications.

8.3.1   A CDP Definition

The Storage Networking Industry Association (SNIA) defines CDP as 'a methodology that continuously captures or tracks data modifications and stores changes independent of the primary data, enabling recovery points from any point in the past. CDP systems may be block-, file- or application-based and can provide fine granularities of restorable objects to infinitely variable recovery points.'

To comply with this definition, applications that offer CDP need to offer a mechanism that allows data written to the primary storage to be continuously captured in a location separate from the primary data and provide any point in time recovery capability. Accordingly, some CDP applications where the updates are happening on regular intervals instead of being triggered by the changes are labelled 'near' CDP.

This is an important distinction because the most significant change that CDP offers over the traditional data protection method is that although the latter requires a schedule to be set for the backup to happen, the former does not require the backup to be scheduled as the data is being backed up continuously. This is where the advocates of CDP stress that by removing the scheduling part, a significant component (the scheduler) of the traditional data protection application is no longer needed. This makes the application simpler to develop, maintain and most importantly use and manage. Additionally, now the backup administrator does not have to struggle to contain the backup activities within the backup window. However, the sceptics would argue that removing the scheduled backup does not eliminate the system resources required to perform the backup. The only difference is that now the tax has been spread out through the day instead of being confined to a specific backup window controlled by the backup administrator. Additionally, this tax could be infringing on the system resources of the business application the server is designed to run.

Before we dive deeper into the benefits and drawbacks of CDP, let us look at some theoretical implementations of this technology. In theory, CDP is a merger between data protection and replication. The use of replication technology plus some basic essentials to allow the user to select a granular RPO provides the essential ingredients for CDP. A derivation of this is the addition of snapshot technology to the mix to enhance performance, albeit reducing the granularity of the recovery RPO from any point in time to a few points during a 24 h period. For the purposes of this discussion, we will divide CDP applications into two major categories: applications that make heavy use of replication at a very granular level and applications that are near CDP and make use of snapshots for periodic updates and refresh of the backup copy.

8.3.2 CDP Using Byte Level Replication

A very basic implementation of CDP is for the application to keep track, on separate storage from the primary storage, of every I/O operation on the primary data. In other words, when a write operation happens on the primary data, the operation and all its parameters are copied to the secondary storage site. For example, let us say that a certain business application at time t_1 requested changing block x located in location l_1 with block x'. This operation is executed on the primary data, and the same commands are copied to a journal file on the secondary storage maintained by the CDP application. Then at t_2 the business application requests the OS to change block y to y' in location l_2. The details of the second operation are copied and added to the journal on the secondary storage maintained by the CDP application.

The captured and saved data can be used for restores in one of the two ways:

- Applied in reverse order at t_3, we can bring the data back to its state of t_0 or right before t_1 was applied. In fact, we have the ability to decide if we want to go back all the way to t_0 or if our RPO should be right after t_1 was applied.
- If at t_0 we had on the secondary site a replica of the primary site, we could take the replica and apply the first operation if our RPO was t_1 or apply both of them and obtain the t_2 state.

In this implementation, the user has the opportunity to choose from a very granular list of RPOs. Although this may be desirable, it is probably not required or sought in most cases. Additionally, the more granular the recovery, the more difficult the management and the worse the performance. Therefore, a more practical way would be to accumulate a certain amount of changes in a journal on the primary storage and periodically replicate this journal. The frequency can be determined by the size of the journal or time. If the application offers a mechanism to adjust the frequency to a desired level, then the user can adjust it to the needs of the business application to meet the SLA. In this case, in one data center using the same CDP application, the data protection administrator can provide multiple levels of RPO to his or her customers depending on their needs.

8.3.3   CDP or 'Near' CDP Using Snapshot Technology

A more common implementation involves the integration of both snapshots and replication with data protection. In this method, the snapshots are replicated to the secondary storage instead of the individual operation or a journal of multiple operations. In other words, on a regular basis – 1 hour or 3 hours or even 24 hours – the CDP application agent on the primary server takes a snapshot of the data and replicates the snapshot containing the changes to the secondary storage.

Some may argue that this kind of granularity is good enough, but the purist would not consider this to be a true CDP application and would call it a 'near' CDP application. In a true CDP implementation, the user can theoretically go back to any point in time. This means that if the user has identified which write or update corrupted the data, he or she can recover up to that point and stop at the corrupting write or update.

However, with the snapshot approach, the granularity of the RPO is limited to the frequency at which these snapshots were taken.

For example, if the snapshots are being taken every 8 h: noon, 8 p.m. and 4 a.m., a virus attacked the system at 11 a.m., the presence of the virus was discovered at 2 p.m. and the user decides to do a recovery, then the noon snapshot is of no use and the only recourse is the one taken at 4 a.m. So in this scenario all the data generated since 4 a.m. would have to be sacrificed or recreated.

Another differentiating implementation parameter is whether the application is purely a software application or an appliance. Some vendors have been offering a software-only CDP application whereas others have seen a better proposition value for their customers in offering them an entire solution with hardware included. The appliance model reduces the development, testing and support costs on the vendor but may not necessarily pass on those cost savings to the customer.

8.3.4 Benefits and Technical Considerations of CDP

CDP offers many unique capabilities that are not found in the traditional backup and recovery paradigm. These capabilities are so unique that some analysts are calling CDP the most significant event in the backup market in the last 10 years. The fact that the process of backing up the data is continuous makes the scheduling aspect of data protection and dealing with the backup window headaches a thing of the past. The potential to offer a significantly higher RTO and an RPO so granular that it can bring data back from a few minutes ago could save businesses a great deal of money on what otherwise would be wasted productivity and lost orders. This is particularity true in situations where recovery is done following a virus attack or a data corruption incident.

CDP applications are beginning to offer solutions to deal with problems that were in the past relegated to RAID (redundant array of independent (inexpensive) disks) systems, mirroring and replication. In fact, CDP applications combine many of the benefits provided by hardware through less costly software solutions.

Because this technology is purely based on disk, application vendors have the chance to equip their applications with end-user restore functionality. This offers their customers a data protection solution that not only has excellent RPO and RTO but also allows the end users to restore their own data. This can reduce the cost associated with the restores conducted by the backup administrator and free them to do more of the planning and management work.

However, like any new technology, CDP is not free of limitations and challenges. Most CDP applications today do not offer a good way or any way to move the data copied by CDP to tape for longer retention. In fact, these applications do not offer the tiering that is often provided by traditional data protection applications. Organizations that have to retain data of 7 or more years for compliance reasons will find it costly to store all that data online, and the application may have performance issues trying to catalogue and manage this large amount of data.

Another limitation visible in some CDP applications today is business application awareness. Traditional tape-based applications have had enough time to develop database and email application agents that take advantage of some of the database built in tools to provide faster backups. The CDP applications available today have yet to provide a comprehensive offering of business application agents and a proposition value for these agents to their customers and to the users of CDP in general.

The true benefits and return on investment (ROI) value of these CDP applications are not going to be fully visible until features like end-user recovery and the ability to manage the granularity of the RPO and the versioning of data are provided. Although the adoption rates for CDP applications are healthy and the forecast is positive, the market is still small and some technical limitations that may be outside of the control of the CDP application developers still need to be overcome. For example, those CDP applications that use snapshots need to make sure that the OS they are running on can support the frequencies of snapshots desired. In other words, if the application provides the user the ability to take snapshots every minute but the OS can only accommodate one snapshot every 10 min, then this would be a limitation.

8.4   DATA REDUCTION: SIS

The Internet and the new communication technologies paired with affordable computing power have played a significant role in the phenomenal growth of data in the last decade. Furthermore, the rapid adoption of these new communication tools has permanently changed the business infrastructure and business processes around the world.

8.4.1   Primary Data Growth and Secondary Data Explosion

Both at home and at the workplace, the handwritten letters sent using postal services and the typewritten memorandums have been replaced by multiple email, voice mail, instant messenger and other types of applications that provide reliable and rapid communication means for personal and business use. This growth in the adoption of email is captured in a study that estimated the growth of email messages sent around the world from 31 billion in 2003 to 62 billion in 2006.

The new communication technologies paired with the abundance of computing power at home and in the office have created new habits. Many users are storing large amounts of electronic data containing songs, pictures and video clips they have downloaded from the Internet or home produced using personal photo and video instruments. At work, in addition to the use of email and voice mail, the growing use of portals to share data amongst co-workers, blogs to communicate internally and externally, and the heavy reliance on office tools for day-to-day productivity continue to fuel this data growth.

To give you an idea how massive this data growth is, we cite a recent study conducted by the School of Information Management and Systems at the University of California at Berkley, which estimated that the world produces between one and two exabytes of unique information per year, roughly equivalent to 250 megabytes (MB) for every man, woman and child on this planet. From the data protection point of view, this growth of primary data can only mean an exponential growth of the secondary data for two reasons:

- Multiple copies of the primary data may need to be retained for restore and disaster recovery purposes.
- Additional copies may need to be retained or archived to meet compliance requirements.

This creates a serious problem for IT organizations in general and data protection administrators in particular. Although the cost of consuming disk for storing primary data is justified by the benefits that data provides to the company, it is often difficult to totally justify the need for the additional cost incurred for protecting that data and meeting compliance requirements. What may seem a good reason to the data protection administrator for keeping multiple copies of the same data for rapid recovery and compliance often seems unnecessary and non-practical to others in the organization.

8.4.2   Issues With Today's Secondary Data Storage

Using today's standard data protection and archiving processes, the retention of multiple copies of files, records or data objects is sometimes necessary to meet the restore SLA. Although this redundancy may help with rapid restores, it can also create a number of issues. Multiple sets of backups may contain the same data object that has not changed over a long period of time. For example, an educational institution that retains full backup sets of every week and daily incremental backups for an entire semester is retaining 14–16 sets of data. This data is retained to obtain an RTO granularity of any day in the semester. However, the data sets will most certainly have a number of identical data objects stored 10 or more times.

   Another issue is the case of a single email that has a large audience. Take for example the situation where the CEO of a company sends an announcement with an attachment in the form of an email to all employees. There are as many number of these identical emails as there are employees in the company. Assuming that everybody stores his or her emails separately and all employee mailboxes are being backed up, the company and the storage resources of the company will likely end up with thousands or even tens of thousands of copies of the same email and attachment. What is even worse is that when the employee email boxes are archived, all the instances may get archived as separate data objects.

Tales From the Real World

According to a number of IT professionals the authors have talked to, who are responsible for data protection in their companies, the most acute problem is the lack of knowledge of the number of copies they have retained as a result of the data protection and archiving process. This not only consuming additional space but is also a liability. They fear that when this data is no longer needed for data protection and has satisfied the legal retention criteria for compliance, it needs to be deleted. However, if all the copies are not deleted due to the lack of awareness of their existence and some of them are discovered later during an audit, it can cause legal problems for the business.

8.4.3   Growth of the Geographically Dispersed Business Model

In addition to the issue of rapid data growth, the speed and reliability in communication changed the way corporations do business around the world. The need to have all employees in close quarters to drive collaboration and enhance communication is no longer essential when new electronic communication technologies are available. More and more people are telecommuting, and companies that are merging or acquiring other companies are not considering consolidation of location as essential as they did before.

Therefore, a more geographically dispersed model is becoming common. The same communication technologies that have facilitated the creation of these models are providing features that are specifically designed to support and promote communications for geographically dispersed companies across the country and around the world. A common model today for an insurance company, bank or retail chain is to centralize IT into one, two or few data centers and support the rest of the offices through these data centers. This model is not limited to service-oriented business but is becoming more and more pervasive in other industries such as manufacturing and even healthcare.

Although basic IT services may have been centralized, data protection remains behind. An analyst survey of data protection professionals showed that 60 % of the respondents indicated the lack of adequate remote office data protection solutions or dissatisfaction with their existing solution. To back up remote office data from the data center over the wide area network (WAN) is expensive. Furthermore, to connect all remote offices of a corporation to the WAN for data protection, especially when these remote offices are small, may not be practical. This type of situation leads to one of two ways:

- expensive data protection processes
- no data protection at all

8.4.4   Issues with Remote Office Backups in the Traditional Data Protection Model

Many customers have implemented media servers fully equipped with tape drives and adequate supplies of tapes in the remote office. They hope that by investing in the remote office, they could provide better protection to their remote offices and maintain the integrity of the entire corporation's data. In spite of their investments, quite often they may not end up with the desired results.

Unless these remote offices are large enough to require a full-time IT person to be on-site, the IT department will likely rely on an administrative staff person to take care of human tasks for data protection. For example, the receptionist or an administrative assistant will be asked to insert a tape at the end of the day and rotate the tapes based on a schedule. Although this process seems reasonable, it is proven to be unreliable causing a false sense of security, whereas in reality when a recovery is needed, the actual data may not exist.

Tales From the Real World

Over the years in talking to IT professionals who implemented these types of processes, the authors had the opportunity to hear some very interesting stories. In one case, the person who was responsible for taking the backup set out of the office (vaulting) was asked where he kept the tapes. Interestingly enough, the response was in his car. In another case, we were told that during an audit it was discovered that the person in the remote office who was responsible for inserting the tape at the end of the day had on a number of occasions inserted the wrong tape in the tape drive. Although keeping the tape in the car may provide the physical separation of the primary data from the secondary data, it does not guarantee the availability of the secondary data in all instances or most instances. The tape in the car, for example, may be subjected to temperatures beyond what it can handle, and this may damage it. Furthermore, if the car is lost or stolen, the data has to be considered compromised.

Another problem is the amount of bandwidth the remote office would consume if a full backup was conducted from the data center. With the rapid growth of the primary data in the remote office, the bandwidth tax that a full backup over the WAN would impose on the company is substantial. Eventually, either the financial cost or the continuous technical challenge of upgrading bandwidth to the remote office would call this method into question.

8.4.5   SIS as a Solution to Remote Office and Data Redundancy

To address the problem of data growth and the ensuing need to use an ever-increasing amount of disk space to protect the data for recovery and archive it for compliance, a practical solution is needed, a solution that provides the user with the ability to store backup data in a space optimized way. Consequently, if such a solution exists, it would solve the problem of providing an adequate data protection solution for the remote office and significantly reduce the cost of data protection.

The concept in itself is simple; if the redundancy can be identified upfront, only one instance of it would need to be stored in the secondary storage. The derived benefit of this identification would also help in reducing the amount of data sent over the WAN when backups of the remote office are done from the data center. In the following sections, let us examine various possibilities for the implementation of removing data duplication on the secondary data.

8.4.6   Data Redundancy Elimination Using SIS

The simplest way to eliminate redundant data is to identify identical files or email messages and store only one instance of the data object and build pointers to it from the other instances where the duplicate data has been removed. One could argue that this type of redundant data elimination can be accomplished through disk-based synthetic backups. It is also valid to argue that the reduction in data through this method is minimal because the redundancy reduction is limited to identical files.

A more advanced method tries to remove redundant chunks of data which we will call 'segments' for the rest of this discussion. This approach attempts to eliminate the duplicate data throughout the file system by segmenting the files into multiple equal segments of a certain size determined by the application or the user. Each of the segments is then given a special identifier using a hashing algorithm. These identifiers are stored in a database.

When subsequent backups occur, the process of segmentation and identification is repeated. However, now a comparison is conducted between the new identifiers and the existing ones. Because the identifiers are unique, whenever a match is found, an assumption is made that a similar segment exists and therefore a pointer is build to the segment and an entry is generated in the database.

In this manner not only is redundant data eliminated from the backup of the client that is being backed up but it is also eliminated from the backups of all the clients in that domain. So there is only one instance of that segment stored for all the clients in that data protection domain. Another significant benefit occurs during the backup of a

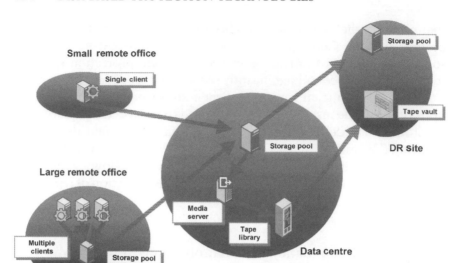

Figure 8.1   An example of the usage of SIS technology in a geographically dispersed enterprise.

remote office when the identifier generated at the remote office has an identical match in the data center.

If the client is at a remote office and the initial backup data is in the data center along with the identifier database, the data protection application agent in the remote office will send only the identifier across the WAN to see if the segment of data that the identifier represents is already there. If it is, a pointer is created. Only when a similar identifier is not found that the entire segment is shipped across the WAN. If the segment is 1 MB and the identifier is 128 bytes, for every identifier found, only 128 bytes are sent instead of the entire MB.

Figure 8.1 shows an example of a couple of remote offices connected to a data center which provides replication to another site as well as the ability to move backup data to tape and transfer the tapes to a vaulting facility. This example explicitly depicts two remote offices to underline the SIS benefit derived within a client, a set of clients in a remote office and within a domain.

8.4.7   Benefits and Technical Considerations of SIS

The advantages of SIS are obvious. The substantial reduction in storage requirements not only reduces the cost of storage required for data protection but also reduces the burden of managing hardware by storing more data in less storage, which occupies less space. Some SIS applications can provide up to 50 times reduction in storage in certain types of data. Furthermore, when the SIS application is designed to reduce redundancy across the entire data protection domain, the benefits are even greater. The amount of data that needs to be sent from a remote office is further reduced, allowing this type of transfer to happen over very low bandwidth lines.

For example, using Figure 8.1 as an illustration, assume that a text document was sent to all users in the company and therefore was found in every machine in the data center and the remote offices. In the large remote office in the illustration above, each of the three servers will have at least one copy of the document. Once a copy is backed up to the large remote office storage pool, the other servers need not send their copies but just the signature or (identifier). When the large remote office storage pool is being backed up to the data center, only one copy is sent. Furthermore, when the small remote office is being backed up to the data center, the document need not be sent but only the signature as it already exists in the data center, and so is the case with any other remote offices that may have that document and are being backed up to the data center. This capability allows remote offices to be backed up with very little use of network bandwidth. Additionally, due to the ability to back up remote offices with minimum bandwidth, users may consider backing up data from laptops and desktops that in the past was not being backed up in fear of clogging the WAN by consuming large amounts of bandwidth.

However, like any new technology there are some considerations that need to be given to the parameters that control the functionality of this technology. One such parameter is the size of the segment. The greater the size of the segment, the fewer the transfers of identifiers accross the WAN. However, with big segments the potential of finding greater numbers of redundant segments declines. Additionally, with large segments, more bandwidth is consumed when a match is not found. Also note that not all data is created equally and some data sets may have more redundancy than others.

Moreover, like all the techniques we have seen in this chapter, this one too relies on disk. Therefore, the storage facility in the data center must be protected either by transferring the data to tape or through replication to another site. Also, users who need to store data on tape for longer retention or archive will have to find a SIS data protection application that provides a way to copy the information from the disk storage to tape.

Some consider the high dependency of multiple data objects on a single segment highly risky. The argument is that if the segment is corrupted or lost for one reason or another, all the data pointing to that segment is no longer valid and usable. Theoretically, this is true but it is no riskier than any other algorithms applied in other technologies such as disk synthetics for example. One way to remedy this is to use the storage pool (Figure 8.1) for rapid recoveries and back it up or replicate it to other media for long-term retention.

Another contention is the possibility of collision. The claim here is that the hashing algorithm may produce two identical keys that in fact represent two different segments. Although mathematically correct, in practicality it would take a several hundred year backup with today's data transfer speeds for this kind of collision to happen, especially when the application is using advanced hashing algorithms such as MD5. Additionally, data protection applications could provide algorithms that detect collisions easily, and concerned users should look for that kind of functionality when considering a SIS data protection application.

One concern recently mentioned is that sharing a segment of data between two or more backup data sets may be a type of security vulnerability. The idea is that the data coming from two separate servers may share one or multiple segments, whereas the two servers may belong to two separate organizations that do not have access to each other's data. Although this may seem like a problem, in reality it is not because that segment of data is meaningless in itself. It does not constitute a coherent set of data without the rest of the segments. Furthermore, access to the segment is restricted by the SIS data protection application's access control mechanism, and only that application can put it together with the rest of the data to recreate it in its original form.

Perhaps, the most serious concern is that although SIS works well for file-based data by removing redundancy, it does not work as well with unique data or business application data like databases. Because of the nature of the data, users may not see the same level of data reduction when backing up databases and other unique types of data as they may notice with file-based data. Another valid concern is the reliance a SIS data protection application may have on its own database. Keeping track of which identifier represents which segment requires a database management system (DBMS). Updating, managing and resolving corruption issues for such a database can add to the backup time and the risk associated with the backup operation.

8.5   NEW PRICING PARADIGMS FOR DISK-BASED PROTECTION

The introduction of disk technology into data centers for data protection and the reliance of technologies like CDP and SIS on disk have invalidated some of the assumptions and proposition values data protection application vendors held with traditional backup and recovery software. Typically, a traditional data protection application is licensed either by machine or by CPU. Additionally, the user may have to license tape drives separately, and the combination of the two constitutes the basic license cost for an enterprise data protection application. Occasionally, the data protection application vendor will offer additional options for a separate price intended to enhance one or more aspects of the application such as database agents, encryption or a snapshot module.

In this model, the user is paying for the software based on a combination of the following parameters:

- The size of their environment or computing capacity, which is measured by the number of machines or CPUs.
- The amount of data that the application is protecting is covered by the licenses for the tape drives.

This kind of combination is intended to provide a fair way to pay for the usage of the data protection application license. Pricing the application by CPU and/or machine alone would be considered by some as unfair to those organizations that have high computing capacities and relatively small amounts of data to protect like a research lab or engineering organization for example. On the contrary, basing the price on the number of tape drives alone would not be viewed fair by those organizations that cannot manage to keep their tape drives busy all the time.

However, with the shift towards disk-based backup and the saturation of the data protection market, many vendors are reconsidering their pricing policies. Additionally, users are looking for predictable cost models that will allow them to budget their yearly software costs just like they can do with hardware costs by predicting their data growth. For these reasons, one of the alternatives data protection application vendors developed was an 'all you can eat' model that they can offer to large organizations where application license management may be challenging and in some cases not feasible anyway without a significant cost. However, offering site licenses poses another difficulty; neither the application vendors nor the users know if they are getting a good deal upfront. No one knows how well either party does on such a deal until after the conclusion or probably after an audit.

Therefore, many price analysts who are dealing with disk-based applications feel that a more fair and predictable model would be the one based on capacity. This kind of model feels good to the users because they are not paying for the application but the usage – after all, data protection applications are data moving applications anyway and the more data you move the more use of the application you make. It also feels good to the vendor because licensing by machine or CPU is certainly not very attractive in a world where people are consolidating servers and CPUs are becoming more and more powerful. Additionally, all storage forecasters are predicting exponential data growth in the years to come. A capacity-based model is likely to continue to pay in growing increments and could be more profitable and certainly more new revenue generating than a server-based model. And for those users who like to budget their software cost, a capacity-based model is a lot more predictable than a server-based one.

So, although in theory a capacity-based pricing model may seem to have the answers for both the vendor and the user, in reality it is not as easy to adopt and implement. In general, adopting new pricing paradigms are not recommended as they are very risky and can cause some serious accounting and finance problems; a new model can disrupt the continuity of financial tracking and therefore make it very difficult to assess the performance of the product and in some cases the company. Some software vendors who have implemented new pricing models have had to revert back to their former models and cause themselves and their customers a great deal of confusion instead of bringing simplicity. So let us look at some of the key pricing elements behind capacity pricing.

8.5.1   Source Versus Target

Basically, the source refers to the primary data and the target refers to the secondary data or the protection copy of the data. So in this section we will discuss the benefits of charging by the capacity of the primary data versus the capacity of the secondary data.

In theory, charging by the capacity of the source may seem beneficial to the application vendor because the revenues derived from the software would scale linearly with the growth of the source capacity and from the customer's point of view, this model would be consistent with the way they buy disk.

However, these two benefits aside, charging based on the source capacity is inconsistent with the traditional model in place in which tape drive usage is licensed. Another item to consider is the retention policy. Normally, a user will retain multiple versions of the source data. If charged by the capacity of the source, the user may not be paying a fair share for the usage of the data protection application, which is being used to create multiple secondary versions of the source data. It may also be unfair to charge by the source for those data centers that have more than one data protection application or may be content protecting some of their data to disk and the rest to tape. It could be difficult to segregate between the various types of data and their targets.

Another factor to consider here is the data reduction or compression. If the secondary data is being reduced using an advanced data reduction technology like SIS, the user may be charged on a capacity that is seriously smaller than what it would have been if the data is not reduced.

8.5.2 Tiered Versus Nontiered

Many software application vendors license their software based on the size of the machine that will run the application. They group machines in 'tiers' and have license prices increasing from lower tiers to the higher tiers. There can be as little as two tiers or as many as four, five or even nine.

This concept is basically implemented so that the cost of the software application is somewhat proportional to the cost of the machine. In the capacity model, the tiers provide a volume discount to the user. For example, a user who is buying 20 terabytes (TB) of capacity is probably going to want to get a cheaper price per TB than the one who is buying only 2 TB.

However, to implement a continuously tiered model is very difficult. In this model, the price is automatically calculated as a function of the desired capacity, taking into consideration a certain discount parameter (Figure 8.2, left). Because of the difficulties associated with managing such a model, many data protection application vendors use a step tiered model (Figure 8.2, right) in which a certain discount is

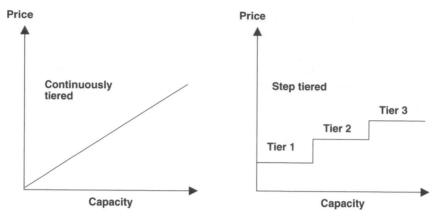

Figure 8.2 Continuously tiered versus step tiered pricing models

applied to the entire tier regardless of whether the capacity purchased is at the beginning or the end of the tier.

Sales people usually prefer a nontiered model because of its simplicity – less stock keeping units (SKU) in general. Additionally, it provides them with a greater deal of flexibility to discount and take into consideration other aspects of the customer background such as previous purchases, loyalty to the vendor and so on. On the contrary, a tiered model has the potential to offer a more consistent volume discount across the entire customer base of a vendor.

8.5.3   Size of the Increments

Another key element in disk-based data protection pricing, especially when dealing with a tiered model, is the size of the increments in capacity. Some data protection application vendors offer increments that match the size of common storage offerings, for example 4 TB, 8 TB and 10 TB options. Others have mostly adopted a decimal system, where the increments are something like 1 TB, 5 TB and 10 TB.

The size of the increments offered by the vendor is important because customers may try to buy what they need today in capacity instead of what they will need tomorrow. However, if they buy too much capacity, they may be licensing storage they are not going to use for a while and even with a large volume discount it may not make sense to buy a large amount of capacity.

The unit used to define the tiers is also important. It probably does not make sense to sell capacity to large data centers with units less than 1 TB. However, organizations that are using disk-based data protection for laptops may prefer a smaller unit such as MB or GB and purchase accordingly.

The other issue with volume discounts is to define if the volume discount is applied to the total capacity purchased since day 1 or only at the time of purchase. In other words, if there is a volume discount at the 5-TB increment and the user has bought 1-TB increments every month, once they buy their fifth 1-TB increment, do they qualify for the discount or do they have to buy all 5 TB all at once to qualify.

8.6   CONCLUSION

In this chapter, we extended the discussion of synthetic backups that we started in the previous chapter to address the concept of disk-based synthetic backup. Applying synthetics in pure disk environments provides some flexibility that is not available when the synthetic engine must deal with tape media. This flexibility can provide for synthetic full backups that use fewer disks and may provide faster recovery.

We also talked about CDP and the benefits and potentials it brings as well as the drawbacks and limitations that normally accompany any new technology that has not been completely adopted by the user community and refined by the vendor. We then discussed another new technology, SIS, which will become more and more visible in data protection applications and in providing data reduction for secondary data storage.

Finally, we talked about the changing pricing paradigms caused by the introduction of disk in data protection and the various impacts the changes in pricing may have on the users as well as the vendors.

Chapter 9

Managing Data Life Cycle and Storage

9.1   INTRODUCTION

From the home user to the data center, disk capacities are growing. Applications are driving data growth and using more and more of the disk capacity that is becoming available to them. Email, video streaming, intranet web services and new paperless applications are just a few of the applications that are influencing this data growth. Not to be outdone, tape capacities have been increasing at a high rate.

Unstructured data consumes lots of storage. The IT professional looks for many ways to manage this unstructured data and create policies and procedures for how to deal with this data. This chapter will explore definitions of the data life cycle and the myriad of issues that surround managing the life cycle of the data.

9.2   ISSUES SURROUNDING DATA LIFE CYCLE

Like a large home with a lot of storage space, nothing seems to be thrown out. As we have the room, why not just save everything. The saved items will need a backup somewhere, so why not save multiple copies? Then, when you need one particular piece of data, it can be like looking for a needle in a haystack. Or, conversely, when you need to purge some type of data, how can you make sure that you

Digital Data Integrity   David Little, Skip Farmer and Oussama El-Hilali

removed all copies? Magnify this by data center sizes and you could have a very big problem on your hands.

In the past, paper was a common form of media. Memos were sent among office workers, and purchase orders were only kept as paper copies. Now, correspondences are mostly done with email and it is considered an official business tool. Purchase orders, checks and other documents are regularly scanned into electronic records that require storage and management over longer periods of time.

Regulations such as Sarbanes-Oxley (SOX) that impose privacy and regulation requirements have recently added to the complexity. These regulations and other business drivers, which include internal security requirements and restrictions, can be interpreted in many different ways depending on which IT group or area within a business may have oversight. The IT group may even be left with the task of evaluating the requirement. In numerous cases, this may include multiple business units that have oversight and these business groups may have conflicting requirements or detailed conditions which may be difficult to put into practice.

Changes in the types of data being stored and locations of this storage, new regulations and business drivers, and the complexity for end users to locate data that they stored a long time ago but now need necessitate some way to manage all of these issues in a type of process. This process is called data life cycle management. The following are some of the reasons for the growing importance of data life cycle management:

- public disclosure laws;
- changes in legal requirements;
- company mergers;
- management of more and more data.

On top of all of these aspects of data life cycle management is security. The data must be kept securely, and only authorized people should have access to the data. This security practice that is added on top of managing the data should not encumber the user from access or make it too difficult for an IT professional to apply business practices to managing the data. For the most part, these security operations should be transparent to those using an application which is back-ended by a data life cycle application.

IT groups are looking to vendors to help meet these challenges. This is extremely challenging for vendors who must develop and integrate data life cycle management applications. How do current data life cycle applications handle security? As user applications and specific environmental concerns add additional security requirements, how are these being addressed by data life cycle applications? These are just two of the more important topics that must be looked at in an attempt to understand where (and if) the process of managing data life cycle is impacted.

9.3 DATA LIFE CYCLE MANAGEMENT

One of the first questions to tackle is what is 'data life cycle management'? At first glance, this may appear to be rather straightforward. You could define this as managing the data from creation time to when the data is removed. Some application or user creates a file on storage, and then, at some point in the future, a determination is made that the file is no longer needed, so it is removed. This entire process is encompassed in a solution that automates the management of the life cycle of the data.

An important piece to the life cycle that is not accounted for solely in evaluating the data by creation and deletion times is business value. The IT professional, specifically the storage or data administrator, cannot determine the business value of the data but is responsible for managing it, implementing best practices and applying policy. The business user, on the contrary, simply expects to know where his or her file or data is when he or she requests it. These two very separate functions can often be at odds and require that a specific structure be applied when managing the data. So now, we have business users who may not understand the complexity of the infrastructure and IT managers who have a quandary when trying to determine which business value to apply to what data and when to apply it.

Applying the end user business values and the administrative requirements together helps us to define data life cycle management:

'The process of managing business data throughout its lifecycle from conception until disposal across different storage media, within the constraints of the business process.' – Symantec

This includes all electronic copies of the data that were made when applying administrative utilities to the data, such as backup. As we know, backup can be done in many different ways containing multiple copies and locations. This in itself helps illustrate the difficulties of data life cycle management.

9.3.1 Hierarchical Storage Management (HSM) as a Space Management Tool

There are fundamentally two ways to apply data life cycle management: archive and space management. Chapter 1 outlined one of these forms of management – HSM – which falls into the category of space management.

Simply put, space management attempts to reduce the amount of online storage being used by moving old or infrequently accessed data to some type of secondary storage which is usually tape. At one time, the disk was very expensive. To keep the pressure off the disk storage, data was moved to secondary storage to reduce the amount of storage needed, thereby reducing the cost. Space management technology in the form of HSM has been available for quite some time now and was employed by IT managers as a response to cost pressures. This allowed end users to save everything regardless of the value of that data.

Moving the data simply to free up space is usually done by applying a rule-based policy that determines when a file was last accessed or created and how long it has gone without being touched or recalled. A metadata file is then left pointing to the location of the file on secondary storage. When the file is removed because it is no longer needed by the end user, the file is then marked as being able to be removed from the secondary storage but is not always physically removed from tape until some type of consolidation is done. The pointer to the file is also removed.

The space management approach has a more limited scope of use, which is handling file servers and moving file data. This can clearly impact its usefulness in data center environments because the use of the application is limited to file servers which may not be a focus of the data center. Let us say that the data we need to manage is contained in a file system. There are generally limited rules that can be applied to this file system as triggers for the movement of the data. Valid rules may be file inactivity or file size. Files may even be excluded based on file type. As the file system is scanned, these defined rules are used and data that meets the criteria specified is then moved. This can present some problems if the rules are disconnected from data business value. The next two sections will look at space management and archive management in more detail.

9.3.1.1   Space Management Example

A data center has a large file system in place that stores scanned documents, user data and photographic images. If all of this data is kept in the same file system, then it will all be moved according to the same rules. However, let us say that user data needs to be kept somewhere that has faster access, so we want to move it to secondary disk storage, and the scanned documents can go right to tape because the expectation of retrieval is not high. We can solve this problem by creating multiple file systems and applying different rules to each of the file systems. So, in this case we would have a minimum of three file systems that have different rules applied to them.

So it seems like we solved the problem of how to get the data to different migration locations. This still assumes that all of the data in each of the file systems has the same requirements. At this point, we start running into problems. If the scanned documents' file system contains documents that are in different formats or different types of data, such as tax records and billing information, they may require different rules to be applied. Tax records, in our example, need to be kept longer than the copy of a billing record. As the file system is scanned on a space management run, the rules cannot differentiate tax documents from billing documents.

There are still many benefits to the traditional HSM approach to data life cycle management. Authors of the *Resilient Enterprise* have a section dedicated to traditional HSM and its benefits. Some of these benefits are more difficult to measure, such as management cost, but other benefits, such as shorter backup window, may be a little clearer. If we migrate less frequently accessed data to secondary storage, it means we have less to back up from primary storage. In later sections of this chapter, we will discuss other types of data life cycle management.

Tales From the Real World

Administrator poll

Lately, it seems that there are more meetings where traditional HSM topics are coming back. When the discussion comes around to migrating the data, administrators discuss disk and tape. Although many of the solutions in the market today can use both as migration targets, there still seems to be a lot of confusion about what was best for their solution. Although most administrators leaned towards disk, I began taking an informal poll about which is easier to manage. I told a dozen

(continued)

Tales From the Real World (continued)

administrators that there are 20 TB of disk or tape for them to manage and asked which they would prefer to manage and which was easier. Almost all of them said that they think tape is easier to manage.

   Some of the reasons they cited were related to their environment. If they already had tape, adding 20 TB more to an existing solution did not appear to change the way they were going to manage the data. If they did not have tape in the environment and wanted to implement disk, they felt that there were a lot more decisions to be made and that managing the disk took more time and very specialized skills and training. Most of the administrators did feel that management did not treat these two scenarios differently even though they were managed differently, and time required to manage was a key factor. In some cases, entirely different groups of people managed disk or tape (I think all the administrators should get a raise).

9.3.2   Archive Management

The second type of data life cycle management is archive management. The archive process moves data to secondary and tertiary storage, which can be:

- disk
- tape
- optical
- DVD
- newer format

It moves the data based on business policies, rules, and may include legal requirements. The archive process may simply make copies of the data based on the aforementioned reasons and not remove, move or change the original file at all. By change I mean altering metadata of the file to show that it had been copied to a separate location or that it was even being managed at all by an outside application.

   Some definitions of archive refer to it as an area used for long-term storage of data that can be backup copies or copies of data that are no longer in use. A number of definitions identified the target area of storage as the archive and not the process of moving the data to the archive. For the purposes of this discussion, archive will be defined as the management of data through its life cycle by applying business values, user values and legal requirements to the data.

   Keeping in mind the types of data life cycle management and the definitions is very important to architecting a successful data life cycle management structure. In the space management section, we discussed how space management – in the form of HSM – was mostly limited to file data. The archive management process is not limited to file data. It is increasingly being applied to specific applications such as Exchange, SharePoint Portal and Instant Messaging. Archive management is getting closer to what data centers need for true data life cycle management.

With the tools available for archive management, we can apply business rules and compliance rules to the data. This leaves us with the following choices of what can happen to the data:

- Data can be stored for predetermined periods.
- Data can be left were it is, consuming storage.
- Data can be moved to another location.

Users set one set of policies and business drivers set another, and these policies will determine the value and lifespan of this data. For example, a team of end users who do order processing may need access to all of the orders for that month. After the 5th of the next month, they consider the previous month's processes completed and no longer need to access it unless there is a specific problem. The end-user process has created a requirement for a policy that says they must have access to the data from each month until after the 5th of the following month. The business driver may say that although the team that processed the

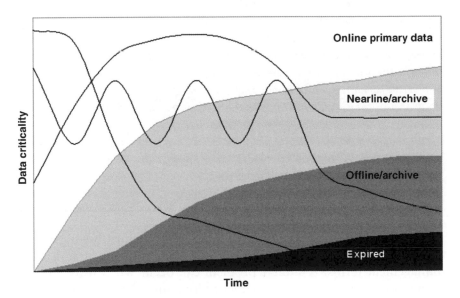

Figure 9.1   Data criticality over time

orders may not need to access the data after a specific time, they will allow their customer access to the data for 3 months. On top of this, there may be legal requirements for how long the company must make these files available or store them. All of these separate pieces contribute to the lifespan of the data.

The criticality of the data may also change over time. Figure 9.1 illustrates some examples of how the data may be accessed over time. Early on, critical data, such as email, will need quick access and should be readily available. Over time this may change. Different types of data will fall into different access categories.

> ## Tales From the Real World Example
>
> ### Archive management example
>
> *I met with a large data center team to determine requirements for protecting and storing data. They were planning to provide a solution where their infrastructure would be a repository for data that they would store for remote offices that they support around the world. Present at the meeting were storage architects, backup administrators, program managers and infrastructure team managers.*
>
> *As we began, high-level architectural diagrams were passed around showing the base components of the design. As I read it I noticed the word archive on one of the systems which was connected to the tape devices. I asked a few questions about business rules and requirements for the data and everyone looked at me a little puzzled. The manager finally said, 'that is just our backup server'. Meanwhile, the backup administrator stated that he was planning to put a backup server in the architecture because he had not seen one on the early diagrams. He said that he thought the archive servers were separate application servers. To make matters worse, the backup administrator had been suggesting to his team manager that they look at procuring a separate tape device for their backup solution as the archive solution had the one in the diagram dedicated to it.*
>
> *What did this tell me? Clearly, the term archive had not been defined and meant two different things to people within the same organization. In very simple terms, the manager was using archive to define backup to tape. The end result could have had a very different outcome from what they intended.*

The archive management example was used to illustrate a problem that can occur when defining backup terminology using data life cycle management terms. A similar problem can occur when using archive management and space management terms interchangeably. Increasingly, I have been working with organizations that have had difficulty in understanding how they can use data life cycle management in their organizations. They know that they have specific requirements and have had trouble articulating that to the vendors who have been suggesting solutions. Understanding that there are currently many definitions to the terms surrounding data life cycle management and that they are not all the same is an important first step. This will make sure that common terms and requirements are clearly defined so that those responsible for coming up with the solution can get the best one to fit the job.

9.3.3   Archive and Space Management Together

Traditional HSM has had limited success in the marketplace. As discussed in earlier sections, this is partly due to limitations imposed by the HSM solution, but some of the limited success may also be historical and simply have a 'fear factor' tied to it.

Legacy HSM solutions were generally stand-alone solutions with a large file server. They relied on older tape hardware, and often the data was moved to the large file server over slower networks. As the HSM was generally not application specific, it was hard to use HSM as application severs began to spread out in a distributed environment.

Administrators felt more comfortable knowing that they had their data accessible. If they could see it online they felt more secure. The 'fear factor' was the 'what if . . .?' factor. What if the tape was bad? What if the retrieval is too slow because so many people need data back? What if I corrupt the file system, can I still retrieve the data? Although in most cases a properly implemented HSM solution can take care of many of the concerns, administrators did not feel comfortable implementing it or saw no reason to implement it when they could continue to ask for disk space or had no specific requirement to implement it.

In general, I think that traditional HSM is a great space saver and the people who implement it see a cost saving and have really seen benefits from the solution. Several of the sites I spoke with about HSM said they had about 25 TB of data on tape for every 2 TB of online storage. Perhaps, perceived complexity of implementing an HSM solution contributed to limited use. Limited use meant that the industry saw this application as a niche application and continued to enhance it based on the limited use rather than expanding the role that space management could play.

What has changed? With more data coming in, applications are driving how the data is accessed and retrieved. Keeping all data online in one place for longer periods increased the space the data was taking online but also made finding specific information more difficult. By combining archive and space management together, the roll of traditional HSM has expanded.

Expanding data capacities has made many servers like the junk drawer in our house. We know what we are looking for is probably in the drawer somewhere but we cannot find it. We spend time pulling things out of the drawer, putting them in other drawers, on the counter or stuffing it back in. There is no organization or management to being able to find it again. You just have to hope you remember where you put it. Giving you a bigger drawer does not mean that you could find something any easier. As a matter of fact, it will probably make it harder and you will just continue to put more junk in it.

Having an application that is smarter about organization of the data and helps the users retrieve data they need and at the same time helps the company or organization manage the data is one of the primary goals of data life cycle management. You can do this with archive management which provides some of the organizational pieces that were missing with HSM, but vendors really need to expand the space management piece to help reorganize the data. This reorganized data can be moved around on the back end to where it is better suited based on infrastructure needs (fast disk, slow disk, tape, etc.), all the while making it easier for people using the data to find and access it.

9.4   APPLICATION CONSIDERATIONS

9.4.1   Email as a Driving Force

A new driving force in business criteria of life cycle management (perhaps the driving force) is email. Email has very quickly become a mission critical form of data that requires 24×7 access. Users store copies of the emails and frequently save emails forever. The volume of email delivered and saved has dramatically increased. With this increase of mail traffic, cost and complexity of managing it have increased as well.

Email as a business tool has introduced a brand new spectrum of issues to be tackled. For example, a leading research organization determined that roughly 50 % of the organizations polled during a 1-year period had no policy to prevent employees from deleting company' important emails. The same research firm found that most companies rely on backups as a preservation tool for email.

Employees are frequently storing emails locally to their system and not on the email servers. Although some administrators attempted to put caps on the amount of storage that users where consuming, the users would get around this by storing mail locally. The problem this introduces is that end users now have company-specific information that could be lost, misused or could just simply take up space that the administrator had originally attempted to free up by introducing storage caps for the end user in the first place.

The content of the email regularly holds attachments. These attachments are getting larger and larger in size and can include video messages, power point documents, excel documents, word documents and even MP3s. Frequently, employees use corporate email for personal and business use which just increases the volume of data traffic. As we saw in Chapter 8, the number of email messages sent worldwide was estimated at 31 billion in 2003. By 2006, this number was expected to double.

All of these issues mean that the email needs to be managed in a way that must allow for quick retrieval of mail information and management of the messages and attachments. Applications that do this must have a way to reach into the email and email attachments and apply some of the data life cycle definitions and policy. Because we have identified email as a driving force, it is generating requirements of its own that must be applied to the data it is sending, receiving and storing.

There are different types of emails and access to email. Some email applications store emails on a server until you download them, some arc accessed through a web interface and still others forward mail on to handheld devices. These are just a few examples of email usage. These types of email usage are continuing to expand the scope of email management.

9.4.2   Instant Messaging

Instant messaging is one of the newer information sources that have entered the data life cycle structure. As email use has increased and become more relied upon, so has instant messaging. As a collaboration tool, instant messaging has become increasingly valuable. If instant messaging is going to be used as a business tool, then most likely we will need to manage the life cycle of the instant messaged data.

Instant messaging is a nice way to work with your peers by giving you the ability to quickly ask a question or share information without having to open an email application, draft an email and send it. You know that you would like to have multiple exchanges with the same person and they are ready to answer you back. An exchange could go something like this:

> **Tales From the Real World**
>
> *Bossyexec: The quarterly numbers did not look so good. Any chance we can add a little spice to them?*
>
> *Financewiz: Sure consider it done!*

This discussion is one that may warrant further investigation with regards to a business rule or compliance practice. So in order to apply some type of rule to these messages, they have to be in the data life cycle construct.

Indirectly, instant messaging is being affected by email policy. Many businesses have published rules around the use of email in the workplace. These same businesses may have in place tools to enforce many of the policies with regards to email. This might include enforcement of content that can be sent and viewed or periodic review of email in the workplace. Instant messaging is seen by some employees as a way around the email policy enforcement. They can send files, data and messages that might have been otherwise blocked or scanned by an email policy. Businesses that allow instant messaging should consider applying the same types of rules and management of the data and content that is being sent with instant messaging systems that are being applied to email.

9.4.3   Business Portals

A portal is a general way to refer to a web-based front end that is used as an entryway to access data that has been put in place for use by many through a common entry. These portals may categorize data in many different ways to allow access by business partners, vendors, customers and employees. The users can log onto the portal and view specific information that they are authorized to view.

The back end of the portal may have different types of data (file system data, database data or other stored formats) that is stored in many locations. The portal application acts as a wrapper to determine who has access to what data and for how long and retrieves the data for the user. The portal presents an interesting set of problems for the data life cycle management application: it must use the business rules that the portal is applying to the data to manage the life cycle. Increasingly, data life cycle applications are being used to interact with portals to help apply policy, storage and retrieval mechanisms for the data.

9.4.4 Applying an Application Strategy

Different applications, such as email, are now driving new technologies and strategies for data life cycle management. As we discussed in the previous several sections, there are other applications that are now being incorporated into these management structures. What we can glean from this is that there is a potential for new applications to follow the ones I have already mentioned into the process flow.

Figure 9.2 gives us an illustration of the strategic approach for an application. As the illustration shows, there is a component to develop a strategy to manage data that usually crosses departments within an organization. Each department or group can set up policies within their own sphere of operations, but to effectively manage the data across the entire organization there must be some way to communicate across all the business units. Technology that enables this type of communication can help mitigate the risk that one area

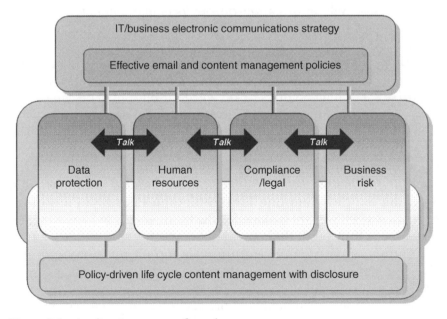

Figure 9.2 Application strategy flow chart

of the organization compromises compliance with the business practice by failing to adhere to policy.

The workflow can be thought of as a practice to help achieve compliance to data management rules that are put in place across all of the business areas. Technology put into place must be able to evaluate these different data management rules, policies and guidelines that are applied to the data. Interpretation of these business rules may vary widely, but having a process to help define and implement any application strategy will be important to determine how the data is managed over its life cycle.

9.4.5 Content Indexing

Many of us are very familiar with Internet search engines. We use them to help find information that we are looking for on the web. The search engine keeps databases of web site information and content to allow you to search terms, words and phrases. A significant element to the search engine is that it provides keyword searches. You type in a word and the search engine looks for relevant entries and displays them back to you. Sometimes when we search for data too many entries are returned. Finding relevant entries then becomes difficult.

As we stockpile information on our personal computers, finding information can become harder. We might remember that we need to recall a document containing some type of information but not what we called it or where we put it. New technology has brought indexing to the desktop. An application crawls our files and creates a small database containing pointers to keywords or phrases. You can search files, emails, instant messenger conversations and more.

Similarly, data life cycle applications can use content indexing. The data life cycle applications can use context indexing engines to store keyword information from many different sources such as some of the applications that we have already discussed. This is a valuable tool in the data life cycle framework. This enables valuable search technology to be used when you are looking for the file or email, and you can only remember a little bit of the content about it.

Context indexing grants end users important functions and also gives businesses necessary tools. It is one of the methods that can be used on the back end of the data life cycle application to help apply business compliance practices. For example, an organization may want to monitor all types of communications and files for sexually explicit or offensive material. They can use the indexing capabilities to monitor for specific words or combinations of words that are being sent between employees. This will then flag these referenced items for further review by human resources or another designated reviewing body.

When applying content indexing to data life cycle management, evaluation of the content engine requirement before implementations should be assessed. Content indexing is not something that is just either on or off. There are various features that different content indexing applications have available. The types of features implemented impact the size of the index.

9.5   ADDITIONAL CONSIDERATIONS

9.5.1   File System Intelligence

File systems have recently begun offering a type of quality storage service. Administrators are able to create file systems that span different types of storage, usually a higher performing, more expensive storage and one that is lower performing and less expensive. The data stored on the file system can then be moved, based on some criteria, from the more expensive primary disk to the less expensive secondary disk. This movement is seamless to the end user or application. This quality of storage service is basically space management applied at the file system level.

Employing business values and legal compliance to this data on the file system is difficult. Although we now have a way to automatically move structured data between more expensive and less expensive hardware, there is problem that traditional HSM has which is applying a model to interpret what the data's relevance is to the company.

9.5.2   File Blocking

System administrators are constantly monitoring the amount of storage that data is consuming. Often, when file systems edge towards a high percentage of used space, administrators comb the data looking for files that can be removed. Many of the files should probably not have been stored in the first place. Managing the data after the file has been written introduces potential problems.

Once the file has been written to disk, there is a good probability that the file will be backed up (potentially more than once) and that the file may be archived. Both of these situations increase the amount of space that this unwanted data has taken up, as well as the fact that it may now have to be managed in the archive stream. The unwanted data could be a file type that the end user was not supposed to store, such as family photos or local PST files that they were using to circumvent space restrictions on the email server.

A tool that can manage data prior to a file being written or stop the data from being archived has the promise of reducing the amount of space being consumed by this superfluous data. Let us consider music files, specifically in the MP3 format. Many companies have policies that restrict personnel from storing MP3s on servers or company-owned desktops and laptops. These policies are frequently difficult to enforce or are sometimes not enforced at all. That means that one MP3 could be stored, backed up and have a disaster recovery copy made of it. That single file has now tripled that amount of space that it was taking up.

Many archive tools have the ability to exclude files by extension. The file may not be archived now but it may still be backed up as we saw in the previous MP3 example. Although an archive tool may be able to restrict the movement of a specific file name, they do not all have the capability to verify that a file that may have a different name extension, may not, in fact, be a type of unwanted file. Continuing with the MP3 example, a file that is an MP3 file named foo.mp3 could be renamed to foo.txt but still be an MP3 file and may be missed by exclude lists of the archive process.

There may be legal reasons to deny or block certain files from being written. If you are a file-sharing provider, for example, you may not want to provide access to data that you did not have permission to disseminate. This could be MP3 files that you do not own copyrights to or documents that you do not have author permission to make available for public use. This degree of file blocking requires that some tag or information about a file be read before the write is allowed.

The preferred method would be to stop the unwanted files from being written to the disk in the first place. We would like to manage the life cycle of the file by restricting where it gets added into the process. We could further fine tune the process by stopping certain types of files

- from ever being written;
- from ever being backed up;
- from ever being archived.

By being able to do this in different steps along the life cycle of the data, we would be managing by exclusion for certain types of data and therefore save space and create an automated task to handle the process.

9.5.3   Backup Integration

Historically, administrators used stand-alone solutions for backup and data life cycle management. Backup administrators managed their area of responsibility without much visibility into the data life cycle applications and vice versa. Much of the same data that is being accessed by data life cycle applications is being accessed by backup applications. This could cause conflicts in several ways:

- load on the system;
- access and retrieval time latency;
- recall of data triggered by a backup.

In the traditional model, files that were being managed by an HSM application needed to be backed up by an HSM aware application. Failure to do so could cause a recall of every single file being accessed by the backup application. As soon as a file was opened with backup intent, the HSM application would consider that a reading of the data and would trigger the recall. This could be disastrous! In this case, there are two options: first, as previously stated, back up the HSM data with an application that can tell that the file is migrated and will not trigger a recall, or second, do not backup the HSM managed file system at all.

In many cases, administrators opted for the latter and did not backup the HSM-managed file system at all. Loss of the file system would require a rebuild of the metadata that was on the file system. One of the benefits of the traditional HSM application was the reduction in the amount of data in the file system. An integrated backup application means when a backup is done there is less data to back up. The restore process is shorter as well, as in many cases only files containing metadata need to be returned to the file system, not the entire file.

As backup windows for applications continue to shrink, integrating data life cycle applications and backup applications together can ensure that faster backup times are met as there is less data to back up. Ultimately, the backup data should be a part of the data life cycle process. As we mentioned in several sections, business policy may apply to all copies of the data which would include backup copies. Organizations that look to integrate backup fully into a data life cycle management structure should see benefits of managing the data.

9.6   SECURITY

Implementing security on data life cycle solutions poses many unique challenges and risks. Chapter 6 has more details about security within data protection in general. This section provides a glimpse into security issues surrounding data life cycle management. Security for data life cycle applications can be broken down into specific areas that each has their own challenge.

Three main areas for security for data life cycle management are

- administration;
- data retrieval;
- data storage.

A data life cycle product is likely to touch multiple applications within an organization. This may lead to more administrative hands on the data life cycle process. Individuals may only need access to one very specific process within the administrative framework of the data life cycle progression. It is important then for vendors to provide applications that allow for role-based access to the data. This means that administrators or people assigned specific functions can only get into their assigned sections.

People accessing the data would like the data to be easier to search to find what they need (which is one of the goals of the data life cycle application). Making this data retrieval process easier may expose the data retrieval process to inappropriate access.

An application that potentially allows users to find their data wherever it resides in the infrastructure could have the potential to be misused, especially as data that needs to be kept longer based on business or compliance rules is presumably the more important data.

Tales From the Real World

'While past attacks were designed to destroy data, today's attacks are increasingly designed to silently steal data for profit without doing noticeable damage that would alert a user to its presence. In the previous Internet Security Threat Report, Symantec cautioned that malicious code for profit was on the rise, and this trend continued during the second half of 2005. Malicious code threats that could reveal confidential *information rose from 74 percent of the top 50 malicious code samples last period to 80 percent this period.' – Symantec Press Release March 2006.*

9.6.1   Public Disclosure

Many states have imposed reporting rules for companies that lose customer data that they have stored. These rules basically revolve around disclosure to the public about the loss of customer data due to situations ranging from lost tapes to unauthorized use or electronic theft. Within a 1-year period at least three large commercial organizations reported that they lost tapes that contained customer-specific information. This information was in the form of backup data but highlighted the dangers to the public of stored data that is put in transit. Scrutiny of how media is secured followed these reports as well as concern from consumers as to how their information is treated. This in turn creates new requirements or puts pressure for vendors to enhance features that are already present in their software.

   Through this disclosure, the public has been made aware of the loss or inappropriate access of data. Not all states have public disclosure laws and rules, so the public reporting does not represent a full accounting of the problem.

9.6.2   Archive as a Secondary Target

Viruses, worms, spyware and adware are all terms that every computer user is becoming more and more familiar with as time goes on. Computer system intrusions and hacking are popping up on news reports periodically that highlight the danger to online systems. Although many new precautions are being taken to secure online data, the area of archiving is rapidly growing and must be secured as well.

   When you manage the data through the life cycle process, presumably the data life cycle application knows about and may create and track multiple copies of data. The original application that holds the original data must secure against unauthorized access; the data life cycle application must do the same. This means that data that is copied to secondary destinations must be secured. If someone is trying to get access to financial records and is unable to do so when attacking the original file location, that someone may attempt to attack the secondary location of the data which is within the data life cycle application and may be assumed to be less secure. The person trying to get this data does not care what the source was as long as it is the same data.

Many credit cards and credit companies store consumer information for long periods. Although there are rules and regulations about what can be stored and for how long, not all of the organizations have mechanisms in place to ensure that they are compliant. An example of this is a recent announcement of the theft of personal identification numbers of thousands of debit cards. A part of the analysis found that companies may be acquiring certain customer information and storing it by mistake. This data was not in primary databases; it was kept in secondary locations and was available for unscrupulous individuals to access.

9.7   COMPLIANCE

Laws and regulations around electronic data have created quite a stir. Companies and agencies are scrambling to become 'compliant'. Compliancy requirements have hit several areas of data management around

- the amount of data stored;
- the structure around storing the data;
- the time period for which the data is kept;
- the time period around retrieval and release of the data to the requesting sources.

Few companies understand their data retention requirements. This is in large part due to the newness of some of the laws and regulations, and how to interpret them and determine when they apply. This increases the danger of data not being available or accessible when requested simply due to confusion on how long data was to be kept or where it was to be stored. Table 9.1 lists only some of the laws and regulations; there are numerous laws and regulations – perhaps thousands. Many of these can even conflict with each other.

Companies doing business globally have to address laws and regulations of the countries that they do business with. This only adds to the complexity. Interpretation of these rules can vary from agency to agency. In many cases, even if there is an understanding of what standards to impose, not always are there tools available to apply the standards.

Table 9.1 Laws and regulations

US laws
Sarbanes-Oxley Act
Patriot Act
E-FOIA/FOIA
HIPAA
GLBA

US regulations
SEC 17a–x
NASD Rules 3010 & 3110
21 CFR Part II
DOD 5015.2

International laws
EU Data Protection
Financial Services Act
Public Record Offices

To highlight just a few examples of regulations and their retention periods, Figure 9.3 shows four regulations. Each of those regulations has corresponding requirements as to how long data must be kept. Failure to meet these requirements can result in hefty fines or potential lawsuits.

This mad scramble around compliancy should include a search for tools that help business become compliant. Remember, the tools can help business become compliant, but simply installing one tool or another will not make the business compliant. Tools available are just that – tools for compliancy. Proper implementation of the tools and enforcement of the policies and rules can help a business or organization become compliant.

9.7.1 Record Deletion

In normal day-to-day operations, files will be created and also deleted. When some type of life cycle management is implemented, the file may be stored in several locations: in the secondary archive locations and backup copies. In the same way that file creation could result in these multiple copies, file deletion may require the removal of each or all of the copies.

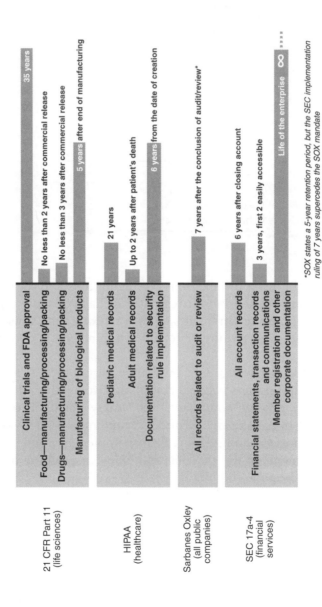

Figure 9.3 Regulation examples

Let us apply two examples to help illustrate the complexities of record deletion. The first example is the storing of legal records that require removal. Expungement is a term used to describe the removal of all information referring to a particular person or incident. In the case of criminal proceedings, data may be stored initially and then require expungement. This means that all copies of the data including backup copies, archive copies and copies in use must be removed. This requires that the administrator knows how many copies are kept and where they are stored.

The second example relates to the end of the life cycle of the data. If your business process has determined that certain data only needs to be kept for 2 years, when that time period is reached, all copies of the data must be removed. If the data is not removed, there may be no legal ramification, but there will certainly be an impact on the administrator who must continue to administer the storage that is being used by this data. There could be a security impact as well. Unscrupulous persons trying to gain certain information may not have to target primary information sources; they could target secondary sources that may be less protected.

Without tools to accomplish these types of record deletion tasks, it again falls on the administrator to apply business logic in determining what files must be removed and make sure that they are removed from the appropriate locations. This can be time consuming and prone to error, such as removal of data that may still be needed.

9.8   CONCLUSION

Old data life cycle management applications have seen resurgence. They are being revitalized by shifting their focus to include more application strategies for data life cycle management. As the speed of tape technologies has increased and the cost of different types of disk technologies has come down, administrators are looking at better ways to manage their data explosion. What may not have been a necessity in the past is looking more and more attractive to solve data management problems.

At the same time, new application requirements are combining with newer technologies for data life cycle management. We have seen in this chapter how new requirements for compliance and an increase in business values are being placed on applications like email. These new requirements and values have resulted in a shift in how data was traditionally managed. Data life cycle management is almost a necessity for any business or agency.

In many cases, businesses are still waiting for a compelling event to determine what action or direction they should take with regards to data life cycle management. The results of this event could be costly. Costs could include expense at implementing a solution rapidly, cost of law suites or cost of fines. Waiting for the compelling event may not be the best way to go. In terms of data life cycle management, a proactive approach is recommended.

Either of these methods (traditional HSM or archive management) must continue to incorporate additional functionality from vendors. Look for tighter integration of backup solutions into the data life cycle management application. There are many applications vying for resource access to a file. Whether it is virus scanning, backup, archive or some other application, vendors need to look for ways to integrate much of this functionality to make sure that performance can be at a maximum.

Organizations and businesses should look into practical ways to limit or manage data on laptops or desktops, or at least make sure that they do not forget about desktops and laptops when creating a data life cycle management strategy. Often, users store data locally on their machines to try to get around a policy that may be applied to their file share or email. Restricting their ability to do this or including their desktop in the strategy will help in the long run.

Chapter 10

Quality Control

10.1 INTRODUCTION

Quality control has been a part of engineering of products for quite sometime. There is extensive engineering and testing done all through the product life cycle. Criteria are created, and a developed product has to meet or exceed the criteria predefined for the product. There are extensive statistical analysis tools, flow charts and processes that are involved in delivering an improved product. A similar practice needs to be applied to data protection and data management - that is, monitoring and managing the quality of the data protection solutions being delivered.

In the past, data protection has generally been solely relegated to the task of backing up and restoring data. Throughout this book, we have discussed ways in which data protection and data management have changed significantly. To encompass a broader range of technologies, business practices and rules and regulations.

This means that the data protection solution, which may already touch most systems in a given environment, now includes a broader array of applications and functions above and beyond the standard backup or data.

Quality control of data protection uses some of the same principles of overall system's quality control. Historically, quality control for data protection was regularly rolled up into the overall system's quality control. There was no real separation of the quality control of the data protection solution itself and the general principles of quality control being applied to the data center. Although the data protection solution

Digital Data Integrity   David Little, Skip Farmer and Oussama El-Hilali

has the widest range of systems that it touched in a data center, it had the narrowest relevance when it came to quality control. With the change in data protection technology and the continuing growing scope of applications and functions supporting data protection, the time has come to apply quality control to data protection on it's own.

Terms like quality assurance, data quality control, total quality management, quality standards and statistical control have been used to describe a process involved in measuring and assuring quality of a delivered product. Each of these types of quality reviews applies a slightly different approach. We will be referring to this process as quality control. Throughout this chapter, we will look at what makes up quality control for data management and how and why these controls should be put into place.

Different applications may make up the data protection suite. Individual controls on management and function of an application may affect the overall success or failure of data protection. We will investigate how these different applications and functions tie into quality control methodology.

10.2   QUALITY CONTROL AS A FRAMEWORK

It is easier to think of quality control as a framework, which has several different moving parts that, when employed together, make up quality control. The framework is made up of different software tools and best practices. There are six major points to quality control:

- design
- define
- manage
- report
- verify
- correct

Design. When designing a data protection solution, quality control should be kept in mind from the beginning. Look at all the steps with regard to quality control that may be relevant to your business, organization or agency and try to include quality control methods into your data protection solution design. This would include tools that may be needed such as host-based software agents or reporting and management tools. Part of the design process can be very tricky. A distinction has to be made as to what may fall into the data protection realm and what may already be included in another data center area that is already covered. If there are multiple types of data protection implementations being done, then each of these areas must fit into the framework design.

For example, if you are using snapshots, bare metal recovery techniques, specific application hot backups, and standard full and incremental backups, all of these must fit into the quality control structure. If you have included only the hot and standard backups, then you will not be able to gather any metrics on the other parts of your process that are being considered types of backups. You would not be able to gather metrics to manage, report, verify or take corrective action on these types of backups.

In many cases, the general result is a disjointed solution when it comes to data protection. Many different teams may be involved all with different goals, expectations and solutions. This can affect the design in both cost and expected results. Cost could be high because many teams are stepping on each other in an attempt to set up their portion of the data protection solution and buying licenses they do not need and managing them through multiple applications, departments or groups within an organization. In worst-case scenarios, this can cause a required recovery process to not be available because it was not understood as being part of the requirement by one of the teams that did only a portion of the backup and recovery solution. In other words, their design did not include a way to recover data that was needed.

Define. Relevant terms, business units and responsible administrative personnel should all be defined early in the process. This step includes the definition of service levels. In some cases, there may be no formal service levels but informal ones probably exist. It is recommended that the informal ones be documented so that all parties are clear on what is expected.

Some sites choose to make these definitions available internally so that everyone is on the same page as to what is expected. One of the focuses here should be on defining recovery time and state for applications. As different technologies are deployed in protecting data, there will be different ways to get the data back. People responsible for the application itself do not always want to know all the details of what is involved in the setup of protection. What they do want to know is that they expect a certain type of recovery to occur in a specified time. Defining all of this up front can avoid the mistake of not being able to meet a recovery time or point and the data protection group saying that the requirement was not asked for therefore it was understood that they would not meet it or debating over what the requirement really was for the data recovery. This discussion should occur before the process is in place so that everyone knows what the expected result will be.

Manage. As the design and definition take place, levels of service will be determined. Manage to these levels of service.

One of the biggest factors regarding management of the levels of service is cost. If the same level of backup and recovery is offered to all applications and end users across the board or service levels are not clearly defined then all users of the data protection solution will request the same level of service. Most of the time they will request the highest level available. As there is no difference in cost to them, they will always ask for the best option available to them.

Although many organizations say they do offer the same level of service across the board there are usually some variations that should be noted. Otherwise, this can put a real strain on administrators who are managing the data protection solution.

Managing to the level of service involves tasks, such as, assigning resources for a particular task:

- Conducting restores or recovery tasks.
- Applying new backup jobs when tasks are submitted for data coming on line.
- Adding more tapes or moving tapes when required.
- Adding additional storage capacity for backups when requested.

As events occur, resources are assigned and tasked based on the definitions from the earlier phase. If everything has the same severity level, then they are assigned on a first problem, first serve basis. There is not really a way to assign a higher priority to any one task. Proper definitions mean that managing of day-to-day operations and tasks as they come can keep productivity high and reduce the effect of reacting to each issue as a high severity issue.

Report. Getting information on the state of your data protection solution is always important. A standard automated reporting process should be implemented.

This would include a way to send reports of varying degrees of granularity to different people that may request them. Accounting for reporting during the design phase can save a lot of time and money.

As there are so many levels of reporting, the design and define steps can help determine what is needed. Many organizations end up having multiple reporting applications in place for the same types of data or several in place because none of them do exactly what is needed. This can be attributed to the difficulty in determining what the actual need was during the design step or the misunderstanding that all backup and restore reporting is the same.

Verify. Once policy is defined, track and verify adherence to the policies or exceptions to the policies. Determine what rules or regulations may be relevant. Then monitor adherence to the compliance standards that have been set forth.

Correct. Allow for corrective action that can continue to improve or change the process as the environment changes. Technology for the data protection landscape has been changing dramatically. The overall flow should allow for changes to be made as new technology is adopted. Correct includes the ability to adapt to changes in environment or requirements.

Not all of the six points (see Figure 10.1) may require action for your environment's quality control. If, for instance, it is determined that there is no governing compliance regulation or law that is applicable, then you may not require a verification of compliance. However, that should be understood upfront and be periodically reviewed during

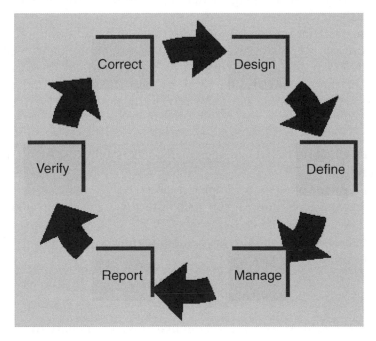

Figure 10.1 Quality control flow.

the corrective step so that if there is a change, compliance can be implemented.

10.3   MANAGING THE SERVICE LEVEL AGREEMENTS (SLAs)

One way to measure quality control is through SLAs, which are agreements between groups or organizations providing a service and users of that service. The agreements define the standards that need to be met. Chapter 5 identified some issues around SLAs and reporting in the backup environment. Here, we will look in more general terms at the overall introduction of SLA metrics into a framework for keeping an eye on quality.

Are SLAs realistic? I ask this because there seems to be a lot of controversy and problems around defining SLAs. Is it possible to create realistic SLAs, which can be delivered, in the data protection arena of information technology? The answer is yes. But this requires careful planning and documenting on the part of both the provider and the user. With all the changes that we have been discussing with regard to data protection, SLA management requires some redefining to measure against these changes.

In the past, data protection was discussed in broad terms as backup and recovery. Reporting was fairly limited, so requirements were not defined very granularly. Chapter 5 mentions some of these criteria. Much of these were around total backup success or total restore success. Here are some examples of criteria that could be measured against:

- Percentage of backup jobs completed successfully (or last job success rate).
- Percentage of restores completed successfully.
- Percentage of time backup and restore system was available for service.
- Help desk and administrative response time.
- Performance of backups.
- Performance of restores.
- Amount of data protected.

In some cases, SLAs were loosely defined or were very hard to measure. This could leave both the provider and the user unhappy, when looking at the measured results. In some cases, SLAs for data protection are a part of sizeable SLAs for file share services or email services. In other words, buried in file share service could be a restore or a recovery time for some portion or the entire file server. Clearly, there is a lot of interpretation that can occur.

Let us look at an example recovery SLA. In some cases, a recovery SLA may state that a restore must be completed within two hours for an application server. Depending on the way the recovery requirements are construed, this could mean several different things:

- recovery of the OS;
- recovery of the application;
- recovery of some portion of the application;
- recovery from bare metal.

Each of these recovery initiatives could be completed in the requisite time frame. However, combinations of these or all of these may not meet the two hours window in this given example. Language of the agreement should specify detailed recovery requirements so that it is clear to all concerned what the expected result would be in the expected time.

10.4   PROTECTION BY BUSINESS UNIT

In many chapters, we have discussed the application of policies, rules or protection based on business units. Many organizations, companies or agencies have separate groups that may get treated differently with regards to data protection. This would be not only management or budget allocation but also reporting. Data protection policies are in a unique position that they often cross business unit boundaries.

As we mentioned previously in Chapter 5, the historical deployment of backup and restore operations tended to take more of a stovepiped approach. Servers would get deployed with individual backup and restore operations and components. Each server may have had individual tape drives or small libraries attached to them. Realizing the cost saving and administrative savings both in terms of cost and time, organizations quickly moved to centralize backup and restore operations.

As new technologies are adopted for data protection, the role of the backup administrator has not as quickly expanded with them. While the traditional backup and recovery methods are now centrally managed, newer technology is sometimes being deployed with individual servers. Snapshot technology is a key example of this. A server being deployed may have specific snapshot requirements. Snapshots are considered by many to be a valid backup. Yet in most cases, backup administrators have very little, if any, visibility into the snapshot functionality. This includes scheduling of the snapshot or initiation of the restore.

In most cases, business units have implemented their own advanced technologies for backup and restore. Deployments of these advanced technologies are starting to look a lot like the traditional model of backup and restore. For instance, a data center has to deploy three Oracle servers, one file server and one mail server. Each of these has different recovery point and time objectives. Each one of these servers at some point has a traditional backup to disk or tape. These servers are administered by the backup and restore team. They handle help desk requests for restores and problems with day-to-day backup that are discovered through standard reporting. Each of these servers implements a snapshot strategy. The Oracle servers each kick off snapshots at a separate time than the file and mail servers. Let us even assume that the three Oracle servers are on different OS platforms.

Although each of the snapshots is used for data recovery, chances are that they are not administered by the backup and restore team. More than likely, each snapshot is handled by an OS administrator and is scheduled individually from system to system. In many cases, these snapshots are being used for worst-case scenarios even though they may be available for individual file recovery because there is no one to administer them or monitor them on a day-to-day basis.

This makes the quality control process very difficult. Bringing the advanced technology deployments into the quality control framework would allow for all the steps to be followed (see Figure 10.1). The business unit can then be applied to the new data protection implementations.

10.4.1 Storage Resource Management (SRM)

There are many different aspects to quality control. SRM is just one component of quality control. SRM tools monitor storage utilization and capacity and provide reports and multiple views on the data they collect. With the increase in disk-based backup technologies, SRM tools are more widely used. Although disk-based solutions are on the rise, tape is still in the infrastructure. Managing to disk at the exclusion of tape drives will only show a part of the picture. Look for an SRM tool that can show both disk and tape.

SRM falls into the manage and report sections of quality control. It allows for the completion of tasks such as

- allocation of storage;
- zoning of SAN components;
- real-time capacity monitoring;
- end-to-end provisioning;
- file level reporting.

These types of tasks allow for operational management of storage resources. Reporting on wasted space and performance allows administrators to have visibility into the environment and provides the capability to make changes to meet SLAs. This, in turn, keeps the quality of the product (data protection and SLA performance) high.

10.5   APPLICATION CONSIDERATIONS

All applications are not created equal. A part of the quality control process is defining what applications will be part of what process. Across all the steps of quality control, applications will require different considerations. It is important to view applications as a part of a total service expected by the users and the business. An end user may access an application through a web front-end service. The back end may contain two database servers, a file server and an authentication server. This particular grouping may necessitate quality control management of the entire segment as a single entity. If any portion of the segment is not performing well or has quality issues, the delivery of the entire service may be affected.

Application performance management is another tool available in the quality control arsenal. Applications that cross multiple tiers can place a collector or listener at each point along the transaction path. Metrics are then gathered and information can be stored in a database. Using correlation capabilities, analysis can be done to determine

- which tier is using the most resources;
- root cause analysis;
- trending;
- predictive analysis.

Application performance management tools provide visibility into and control over application's performance. This enables an organization to provide the highest level of quality and reliability. This is an important service provided by quality control. Data protection solutions that span tiers of service will need to monitor performance and reliability across all tiers. Even if the portions of the data protection solution meet the requirements for some portions of the solution, the overall availability number may be much lower. In other words, if the system is up and available and backups and snapshot occur on time but retrievals are unavailable, the total amount of time the entire solution was available may be well below 100 %.

10.5.1   Corrective Action

A part of quality control is corrective action. There are two types of corrective actions:

1. *Problem response*. Immediate issues and problems will require resolution. Corrective action for these types of problem responses may require patching a system or maintenance of day-to-day operations.
2. *Re-evaluation*. Corrective action can be used in re-evaluating the data protection design or steps. Provisions to allow for changes to the application cycle should be built into the quality control process.

Problem resolution and response are the basic functions that should be available. The most common form is an automated response through a help desk. A notification occurs that there is a problem with an application either through SNMP or other alerting mechanisms. This notification then triggers the generation of a trouble ticket that can be worked on by the appropriate technical response team.

By generating trouble tickets, reporting can be done to measure the response to these tickets, as well as track how many are opened and for what. Knowledge base information can be developed from the areas identified by the trouble tickets. Quick access to self-help tools and pre-emptive changes and adjustments can reduce the overall number of trouble tickets or the mean time to resolution.

The quality control process workflow calls for periodic re-evaluation. On a scheduled basis, the process should be reviewed to determine what changes may be needed. This should be incorporated into quality control from the beginning. The objective is to ensure that the same problems do not re-occur in the future.

10.5.2 Patching

In large environments or data centers that can afford little downtime, there are strict controls in place on what can be installed in production. In general, this refers to whole applications, major updates, patches and may even apply to individual binary files or application changes. This presents an interesting problem to quality control within an environment. Let us look at the pros and cons of patching in production.

Occasionally, we hear a horror story of how a patch or update was put into production and system outage was incurred, causing extensive downtime. Table 10.1 shows just a sample of some of the pros and cons of patching. We certainly want to protect against this worst-case example. With the number of applications being managed increasingly by administrators, they are gradually looking at ways to provide automatic updates or live update functionality.

In one case, I visited a data center that was having several application issues. For quite sometime they have been receiving help desk calls and implementing manual procedures around the issues that they were facing. The increase in the help desk calls was affecting the quality of the product they were delivering. The same issues were being reported multiple times. A known patch update to their application was available but not implemented due to a delay in the patch approval process. The resulting calls to the help desk were affecting their overall rating of the service that was being provided.

Administrators are now looking at the possibility that rather than pushing out updates in all cases, the application agents can perform a pull. This would allow administrators to approve a patch and make it

Table 10.1 Pros and cons of patching

Pros	Cons
Protects against application failure	Causes application failure
Protects against data loss	Causes data loss
Fixes data integrity issue	Causes system outage

available for the application to update when it performs a live update check. At least, an automated policy should be performed to check and verify that systems are at the recommended patch level.

This automated patch verification can ensure that the systems have been updated or identify systems that do not comply with the process. This is especially helpful when there are a large number of systems to be patched and/or multiple shifts of administrators responsible for updating the systems. For example, administrators have identified an issue that is fixed with a specific patch level. The patch goes through the requisite testing or approval process and is then placed on a share from where administrators will perform the update. The updating will be done during a maintenance window by a second set of administrators. This set of administrators goes to the share to run the patch update and selects the wrong directory, which is from the earlier, already deployed patch. They deploy the patch, and the problem they were having still exists but now the systems are out of compliance, as they are actually still at the incorrect patch revision.

Certainly, a verification step would help solve this problem but on a large scale there are a lot for administrators to keep track. A tool that supports the automatic verification of the patch level could be used on this larger scale. In the previous example, an automated policy check would have immediately alerted administrators to the problem, and further inspection could have quickly uncovered the problem.

10.6   POLICY AND COMPLIANCE

A segment of the quality control process is a verification that policy is being adhered to and compliance is being met. Regulation and policy are consuming a larger amount of organization's time and money. Policies will need to be defined and then controlled across the environment. Evidence of compliance must then be kept. Automated review of policy against industry best practices can help lower the cost of compliance.

Many of today's guidelines in organizations are vague, not adhered to or not documented at all. In the define phase of quality control, many of the vague guidelines can be put into actionable policy. A policy management tool then automates the step of assessing adherence to these policies. Regulations such as SOX, HIPAA, FISMA and others can now be included in automated policy examination. An illustration

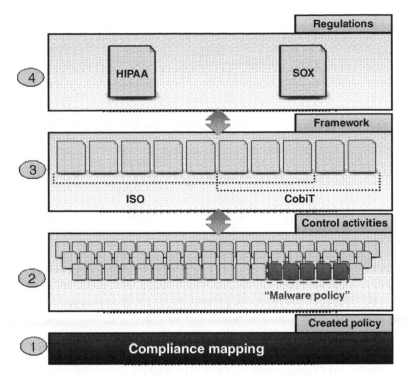

Figure 10.2   Compliance mapping.

for the path of a compliance map is highlighted in figure 10.2. The map starts with the creation of a compliance policy.

Antivirus is a good example of a required compliance check to a policy. Most organizations have a policy that requires that servers put into production have virus protection in place. A policy and compliance tool runs on the system and verifies that the requisite virus scan software and version are installed. If it is not installed, a notification is sent to the appropriate administrator to take corrective action. The same model could be applied to a data protection application that is required prior to deployment of a server.

10.7   COST MODELLING

Once business units, SLAs and reporting structures have been defined, many providers of a service want to bill back for these services that they are providing. In addition, providers of a service or a set of services can adjust their business as technology changes to provide supplementary

Table 10.2   SLA Map

	Bronze	Silver	Gold	Platinum
Frequency (full)	Weekly	Twice/week	Synthesized daily	Low-cost disk + daily synthetic
Frequency (Incremental)	Daily	Twice/day	4 times/day	Hourly
Technique	Cold (application shutdown	Hot backups	Hot backups, snapshot moved off-host	Hot backups, snapshot, instant recovery
Disaster recovery	None	Off-site vault	Offsite vault + local copy	Recovery site, off-site vault + local copy
Restore service	24 hours	4 hours	1 hour	Minutes
Cost	1x	2x	4x	8x

services. With rapidly changing technology, the process should be reviewed to allow for faster deployment of extra services.

Tiered data protection services could offer a broad range of products. Traditional backup and restore from tape through rapid roll back from snapshots can be provided at different service tiers. Strong definitions and design maps will explain what is provided at each tier so that there is no ambiguity. Without these definitions in place, a user could expect rapid roll back of a database, which is at a much higher cost, when all they should be provided are nightly backups to tape.

Table 10.2 illustrates a sample of how different service levels may appear. Distinct techniques for data protection each map to a discrete cost for that level of services. To provide separate levels of service, there should be a way to manage to these levels of service. Accurately measuring the levels of service improves the ability to provide that service and the chances that many different services can be offered. Rather than one offering of backup and restore, you could have four to six levels of backup and restore offerings each with a known cost to both the user and the provider of the service.

Many times I hear from organizations that there is no cost for the different departments that they are providing data protection services for. There may be an internal cost, but really they say there is no way for them to separate the cost of data protection and bill any one group for that cost. Although this may be true, different business units or departments are using different amounts of data protection resources. By using a cost model, a determination can be made as to the 'cost' of data protection resources. This 'cost' could be the administrative time required to resolve problems, resources needed for daily management of systems or even amount of storage utilized. By having measurements of time and resources needed, better accounting can be made of what is required to get a given job done. That way, when department X asks the group responsible for data protection to add 10 new servers to the architecture, they will know how long it will take and what it will cost in terms of resources. The end result will be a more realistic model to complete the tasks.

Billing and chargeback may not be directly a part of the quality control cycle but they are affected by it in the long run. If billing and chargeback are a part of the organization's business model, then quality control will ultimately help by allowing for a better solution to be provided. Helping to meet SLAs, meeting policy and compliance will all add to the reduced cost of deployment and management along with meeting the goals that were set forth to the users of the service. By meeting or exceeding performance metrics, the organization will generate revenue for their services.

10.8   SECURITY

Personnel responsible for quality control will need to have a view into the different areas and the tools that may be required for their day-to-day functions. Their views may cross business units or be restricted to specific business units. In all of these areas, we want to make sure that we give the correct people views they need to have and keep those with prying eyes from the views they should not have.

Integrated quality control tools that permit users to receive defined views across multiple areas help with the security process. If you are using an SRM tool for capacity management and monitoring and a backup reporting tool for operational management and the tools allow for the definition of user functions across both tools, then the user only has to sign in once to receive the appropriate credentials.

As all of the tools may not be integrated, the use of a single sign on tool may be helpful. Single sign on authenticates a user once for multiple applications for a determined session. The user could sign on through a single sign on process that would give him or her view rights in an SRM tool but full administrative rights in a policy management tool.

The benefits of the integrated applications and the single sign on are the reduction of passwords required and the reduction of entry points into the applications. Fewer passwords reduce the chance that a user will take shortcuts that could compromise the password security. Less entry points means that the application entry points that are allowed can be monitored and all users can be directed through a narrower set of these monitored entry points.

When selecting tools or defining your process, look at ways to include audit logging. Audit logging is a practice that records user functions. These functions can be simply who attempts to log in and who successfully logs in to the application. More detailed audit logs include recording of who attempts to make modifications to some operational procedure or policy. Many rules and regulations now require audit logs and trails for specific applications or functions.

Proper practice of security and use of security tools can go a long way. Creating a log in for backup administrator and letting any of the 10 administrators log in with the same user name and password will not allow for individual audit trail logging. It is better to create individual accounts that have similar permissions such as a role of backup administer than to let each of them log in with the same log in. Many applications now allow for integration into (LDAP) active directory or some type of network user database. This way the user can first be authenticated and then given authorization to the specific tools he or she needs.

Applying security patches is a consideration for security as well. As vulnerabilities are discovered in applications, fixes become available. In Section 10.5.2, we discussed the general nature of patching. Security patches often receive quicker approval for deployment in the data center. Similarly to normal patch operations, an automated tool to verify compliance to the patch level is even more critical here. Once a security patch is identified as being required, the policy software can be updated to run a check of all data protection applications to verify that they have the security update. Although today's policy software does not cover all the applications that may be needed, it is certainly headed in that direction.

10.9   CONCLUSION

Providing visibility to end users during the steps of the quality control process grants users assurances that the process is being followed. It presents a kind of check and balances system. The end users can see metrics used and get timely reports as a part of a standard process.

Their having a view into the process ensures that the process is being followed with the agreed upon steps and results.

Quality control is an ongoing process. It is not a one-time check across the data protection method that is put into action when servers are first deployed. Ongoing changes and updates to the environment and applications require the quality control procedure to be continuous. Automated tools are a must for a viable quality control process. Today's quality control process has to take into account not only the applications being deployed but also their compliance to policy and laws and regulations. A combination of policy control, SLA management and resource management works to fulfil the desire of completing the requirement for quality control.

Chapter 11

Tools for the System

11.1 INTRODUCTION

As data protection and recovery have grown in scope and morphed over the years, additional tools have become available to help with aspects of the data protection infrastructure itself. As more servers and data are being protected by a data recovery solution, it is no longer acceptable for the servers responsible for data recovery to be offline. Just a few short years ago, it may have been the standard practice for the data protection solution to be configured without high availability (HA). Having the server offline for maintenance for periods of time was deemed an acceptable position to take.

Tools like HA and provisioning, while widely used in the data center, are increasingly being used for data protection solution components or are a part of the design into how data protection will be managed. This chapter will explain how some of these tools are being put into practice in the sphere of data protection.

11.2 HA

The concept of HA normally refers to systems that have the ability to remain operational 100 % of the time. This is accomplished through built-in redundancy in the system. Clusters are one way to provide HA of business critical applications and data to users. There are a number of clustering applications in the market that offer this type of functionality, but they all usually work in a similar manner.

Digital Data Integrity   David Little, Skip Farmer and Oussama El-Hilali

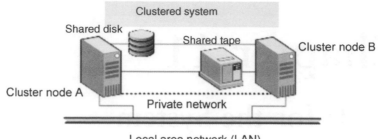

Figure 11.1   Basic two-node cluster.

In a cluster, two or more servers (called *nodes*) are linked in a network and run cluster software. Disk storage is shared between these nodes. Each node has access and uses the shared disks. If a node becomes unavailable, cluster resources migrate to an available node (this is called *failover*). The shared disks and the virtual server are kept available. During the failover, the application and the users experience only a short interruption in service. Figure 11.1 shows a two-node cluster with the basic clustering components: shared storage, private network(s), public network and shared media.

Although HA is primarily intended for business critical applications, many data protection users take advantage of the presence of the clustering application and use it to ensure HA of their data protection application. Many data protection applications run in clustered environments and can provide the redundancy necessary to ensure that backups are done and the ability to do recovery is always available.

There are two separate types of implementations that we will discuss in this section:

- Protecting data that is part of a cluster.
- Clustering a data protection application so that it can be highly available.

11.2.1 Protecting Data that is Part of a Cluster

Quite often a business critical application is set up in a cluster and needs to be highly available. There are several ways for a data protection application to protect the data in a cluster environment. One simple way is to back up the data from a cluster across the network to a separate server running the data protection application by installing the data protection agent on the cluster nodes. This is the easiest method to set up and maintain as the data protection application specific configuration tasks for tape devices, media and so on are kept separate from the set-up and maintenance of the cluster itself.

11.2.2 Clustering a Data Protection Application so that It can be Highly Available

More often an information technology (IT) organization needs to have a highly available data protection plan to ensure that daily backups are happening. This type of requirement usually comes from internal service level agreement or audit where the IT organization needs to prove to the auditors that backups are being done on a regular basis. In these cases, the data protection administrator has to find a way to make the data protection application highly available. This can be done in a couple of ways:

- Use the internal mechanisms of the data protection application and achieve some HA but not full HA or what is called the five nines: '99.999'.
- Use clustering software to accomplish HA for the data protection application.

Sites will generally have some type of administrative server or master server as the brain behind the data protection solution. This master server stores the catalogue of what has been backed up when and location of the backups and handles other tasks such as scheduling and initiation of restores or recoveries. The data protection solution may contain one or many servers that handle the workload of actually moving the data to the backup storage device selected. We are going to refer to these workload servers as media servers.

Some data protection applications that use a client/server architecture will offer a means to fail over the load of one of the media servers to another in case one of the media servers is down. With the appropriate set-up, if one of the media servers is down, the other media server(s) can pick up the backup. These results can also be achieved in some cases through a manual process when the data protection application does not offer this functionality.

To obtain higher levels of availability, some users will use a third-party clustering solution either provided by the OS vendor or the data protection vendor. This clustering solution, in addition to providing HA to the business critical solution, provides HA to the data protection solution. There are a number of ways and configurations that can be used to achieve data protection HA. One involves clustering the media server and the other involves the master server.

Installing a single media server on the cluster as a virtual server application allows the media server to fail over from one of the nodes. This is called a failover of the media server. In some applications, this is done by assigning the data protection application a network name resource (the virtual server name), an IP address resource and a disk resource. The network name and IP address must be unique and set up in your name resolution configuration prior to setting up the failover server.

The virtual server name is used when configuring the media server and any clients to be backed up through this media server will refer to it by its virtual name. When a failover occurs, backup jobs that were running are rescheduled by using the retry logic for a failed backup. The services are then restarted on another node and the backup and restore processing resumes.

Yet another approach is to install the master server on the cluster as a virtual server application. This is called a failover of the master server and as the name implies allows the duties of the master server to fail over from one of the nodes to the other if a failure occurs on the active node. As with the failover of the media server, when a failover master is installed, a network name resource (the virtual server name), an IP address resource and a disk resource are assigned to it. The network name and IP address must be unique and set up in your name resolution configuration prior to setting up the failover server.

The virtual server name is used as the name of the master server for all media servers and clients using this master server. As above, when a failover occurs, backup jobs that were running are rescheduled using the retry logic for a failed backup. Both types of failover servers, master and media, operate in an active/passive failover configuration. The active node and the passive (or failover) node must be of the same type of server, master or media.

11.3   PROVISIONING

An additional tool that can be made available to data protection administrators is a provisioning tool. Provisioning allows for the rapid deployment of new servers or the quick redeployment of servers that already exist. So how can this tool add value in a data protection environment?

There are two areas where provisioning can help:

- growing environments;
- moving from test to production.

11.3.1   Growing Environments

As more servers are added to a given environment, the use of additional servers as a part of the data protection solution may be needed. These servers are the ones that would be used to move data to the storage devices selected. Traditionally, this would mean that new server hardware would be selected and added to the given environment. Then, an operating system would be installed. Patches and network configuration would be done, and then the enterprise data protection application would be installed and configured.

Provisioning can automate the process of installing the operating system, configuring the application and modify network settings as required. In fast growing environments, this speedy deployment enables backup load to be shifted quickly.

Let us look at an example. Let us say that a given data protection deployment was designed to handle 20 TB of data within in a given backup window and scaled to allow 10 % growth or the data being backed up year over year. The servers moving the data to the storage devices can handle the current load and allow for some growth. However, now the organization merges with another one, and the amount of data that it needs to handle is double what it had before.

Although allowing for some growth, clearly the doubling of the data will tax the resources that are currently available to the designed solution. New servers need to be deployed to handle the load. Hardware and storage need to be assigned. In many cases, hardware and resources may actually be available as a part of the acquisition or consolidation of departments. Now, it is just a matter of taking all the steps to get the hardware ready to be used.

Provisioning is becoming increasingly useful especially with blade devices. The data protection components that move the data to the storage devices are designed to handle a specific load. As the load increases, there is a predetermined point where a new blade system gets deployed and is ready to handle a share of the new load. The use of provisioning speeds up this entire process and adds to the scalability of the design.

11.3.2   From Test to Production

Organizations generally like to create test and development systems for applications prior to their utilization in production. The same applies to the pieces and parts that make up the data protection solution. Once the solution is tested and developed, it gets reinstalled and configured in production. The goal, of course, is that the installation and configuration of the data protection solution in production should match how the solution was tested and put together. This usually involves lengthy procedures that should be followed to the latter in production. All of this, of course, takes time and can be prone to user error.

A provisioning solution helps to ensure that the configuration that was finalized during the development and subsequent testing periods is the same configuration that gets deployed in production. Not only would it be the same but it would also be deployed in a reduced period of time.

Provisioning should not be confused with bare metal restore that was discussed in Chapter 4. Bare metal restore is a process designed around rapid recovery of a server. There are a lot of different facets to bare metal restoration that are primarily focused around quickly getting a server back into production with automated recovery tools. Provisioning, on the contrary, is about rapidly deploying or redeploying system resources that may have not even been backed up yet.

11.4   VIRTUALIZATION

Virtualization is the term used to describe something that is simulated but not physically present. In the technology space, there are many types of virtualization. In this section, virtualization will be used to refer to a layer that is the creation of a virtual operating system, one that can use the same underlying physical resources but not be dependent on other virtual operating systems that run on the same physical devices.

In the area of virtualization there are several trends that are occurring: One is doing hardware emulation through the OS level where the host OS runs an application to present emulated hardware on which the guest systems can run. The other is the hardware emulation through firmware. The latter can obtain near-native speeds, whereas the former has limitations (can not emulate 64-bit hardware yet – even on 64-bit host systems).

As applications are placed into these virtual layers, the data protection solution has the responsibility of protecting this data that is self-contained. In some cases, there is the added responsibility of protecting the entire 'virtual image' so that it can be quickly restored. The image would be a collection of files that can be backed up and then restored or moved to the same location or a new location. Similar to a databases, the virtual image needs to be placed into a consistent state before being backed up so that the image can be restored. The host system that runs these virtual operating systems or enclosed images needs to be protected along with the virtual images themselves.

It is generally not recommended for the work horse portions of the backup and recovery solutions to be contained in a virtual image; pieces like the catalogue servers or the media servers. Because of the load that is generated by these servers, they tend to use resources that would not be available on the server. You would tend to run them standalone. If you are trying to maximize the resources on a system and you add virtual images to run applications, you would most likely not have resources available to add the backup server to that system itself and would have to put it on it's own machine.

11.5   SUMMARY

The role of data protection has changed so that it is now an integral piece of the production infrastructure. Like many other applications, this means that it falls prey to some of the same requirements that many other production applications have. These requirements are that the data protection solution is always available. Restores need to be done at anytime and backups, with their ever-shortening windows, need to be done around the clock. There is no longer the luxury of long down-times and outages for the data protection solution.

The solution needs to include tools for quick deployment. Not just new installations but new deployments of the data protection solution components may be required for growth. Quick deployment is required for testing, developing and then finally for implementing. In many data centers, the amount of time it takes to introduce a new media server can be reduced from days to less than 1 hour.

HA and provisioning are often used in production for applications or services that are provided. For some reason, this does not traditionally include data protection components. Using these same tools for the data protection system that are already provided for other areas of the production environment will help keep the backups and restores running and minimize the amount of time it takes to grow and scale a solution.

Conclusion

First, we would like to thank you for allowing us to share our thoughts and views on the past, present and future of data protection and data management. We hope that you have found this to be useful. We started this journey with a look at the historical view of data protection which was primarily based on traditional backup and recovery. We actually took a closer look at the whole process on implementing a backup and recovery system and took a look at some of the goals of this kind of data protection scheme including; being able to enable normal services to resume as quickly as is physically possible after any system component failure or application error; enabling data to be delivered to where it is needed, when it is needed; meeting the regulatory and business data retention requirements; meeting recovery goals and in the event of a disaster, return the business to the required operational level.

We also looked at the steps necessary to achieve these goals which includes making copies of all the data, no matter the type or structure or platform upon which it is stored, or application from which it is born; managing the media that contains these copies and in the case of tape, track the media regardless of the number or location; providing the ability to make additional copies of the data; and being able to scale as the enterprise scales, so that the technology can remain cost-effective.

We drilled down into some of the challenges that are usually encountered when trying to architect this type of backup and recovery systems, even to noting that we really should think of this as a recovery and backup system as you should always architect this kind of system with the recovery requirements as the guiding light.

As this part of the discussion came to a close, we highlighted some of the things that were forcing changes to the way we think of data protection. These changes include the total amount of data in the enterprise; the criticality of data; the complexity of data, from databases, multi-tier applications as well as massive proliferation of unstructured data and rich media content; the complexity of storage infrastructure, including SAN, NAS, DAS, with a lack of standards to enforce consistency in the management of the storage devices; heterogeneous server platforms, including the increased presence of Linux in the production server mix; more aggressive recovery time objectives (RTO) and recovery point objectives (RPO).

Next we took a look at how hierarchical storage management (HSM) can actually be thought of as a part of the data protection and data management scheme. This included looking at how HSM can help to reduce requirements for online storage, reduce file system management, reduce costs of backup media, and reduce management costs.

No discussion of data protection would be complete without taking a look at disaster recovery (DR). Traditionally this has been based on a collection of the backup tapes that are kept at a DR location or at a storage facility. This has led to the development of specific Vaulting solutions that help you to manage the creation and tracking of the backup tapes. Unfortunately in too many cases there was not adequate DR planning or specific training to enable the user to be able to successfully recover from a disaster. This is starting to change. This is one of the areas where we start seeing the merging of data protection and data management tools with the advent of clustering and high availability (HA) solutions that work with the backup and recovery solutions. This is especially driven by the need to better manage RTO and RPO.

There was even an introduction to the evolving solution area of encryption. This is a rapidly growing and changing subject based on business data needs and emerging technologies. We also took a look at overall management and reporting and how that has traditionally been handled. This included a look at service level agreement (SLA) management.

As we moved on we next examined how some of the basic data management tools have started to evolve more towards data protection. This included a look at storage in general and disk management more specifically. This also included a discussion on storage virtualization including definitions of a lot of the basic terms used. We looked at some of the reasons for this virtualization which includes better performance, more availability, overall cost of capacity and manageability. This led to a pretty detailed discussion on RAID.

Another of the disk management tools discussed was replication. The first point made concerning replication was to actually define it and show how it differs from true data protection. Some of the reasons mentioned for using replication are data distribution, data consolidation, off-host processing and DR. As mirroring can provide these same capabilities, why use replication? This question was addressed by highlighting the important factors that differentiate replication from mirroring which are latency, communication reliability and the source-to-target relationship. There was also a discussion on how this technology has been used to set up DR or standby sites.

The logical progression was now to take a look at how backup applications have started integrating some of the data management tools. The most common integration has been the use of snapshots, either mirrors or copy on write (COW) snapshots, as a backup object. These snapshots can also be used as instant recovery objects. Some of the applications have even integrated the newer fast mirror resynchronization techniques. This makes mirrors even more attractive as a backup object. Another of the obvious integration points is backup with replication.

We next took a look at a completely different part of the data protection landscape, BMR or Bare Metal Restore. We took a look at this topic from several different views including first, why consider it at all. A part of this discussion included some of the key points that have hindered the development of a true bare metal restore application. This actually led to a review of the evolution of the bare metal restore capabilities. We looked at some of the native operation system applications that have been developed to try and fill this need. These included IBM's AIX Network Install Manager, HP's HP-UX Ignite-UX, Sun's Solaris Jumpstart, Microsoft's Unattended Text File and Linux Kickstart.

Our journey through the world of BMR included a look at the limitations that the individual operating system solutions faced as well as the challenges that this entire process faced. This included figuring out how to do recoveries to dissimilar disks to completely dissimilar systems. We looked at how one particular application has been able to solve these very tricky problems. Finally we took a peek into the future of BMR. This included a brief discussion on how CDP and SIS are offering their own challenges.

The fast growth of the open systems and the adoption of the Internet in daily business life opened up many opportunities that the mainframe systems could not provide. However, a new set of challenges came and continues to tax this new way of computing including the ability of the open systems to provide the security and data integrity. We saw it fit to dedicate a chapter – chapter 6 – to the issues involving security and briefly touch on encryption which is gaining momentum these days.

The movement of data from one place to another whether is over a network or physically transported on media without compromising its integrity has led to the use of encryption technology in data protection. We also addressed the need of a data protection application to have role-based security so that accessibility to the data can be defined by a clear set of rules and roles. This type of functionally not only allows an application to offer a better system for data protection and access to that data but it also serves a fundamental infrastructure for audit trails. Audit trails are becoming more and more popular with the expansion of regulation and the need to audit data creation and movement activities.

Also addressed was the challenge presented with security vulner-abilities and how it affects applications especially data protection applications. We felt that security issues will continue to dominate the discussions in data protection both from the usage point of view as well as a development topic.

In discussing new features in data protection we covered some of the newer concepts and methods for doing backups and recovery. These concepts included synthetics backups or what is often referred to as incremental forever. A good portion of the discussion on new features included a renewed focus on some of the disk backup methods that can utilize tape or tape-like features including staging and virtual tapes libraries and their expanding role in data protection. The tools and methods are commonly available today in the established data protection applications and can provide users with a great deal of flexibility and cost effectiveness by providing fast restores and reducing the backup window. While we tried to present these technologies in a positive way we made sure to discuss the challenges associated with these technologies so that the user can be aware of them.

Disk-based protection technologies are dedicated to disk-based backup and the new technologies that surrounds a solution based solely on disk. We included a section on continuation of the synthetic backup discussion by describing the effectiveness of synthetic backup done in a purely disk environment. We also talked about continuous data protection and the role it can play in the future to solve the problem of the shrinking backup window while providing fast restore capability. We dedicated a section to single instance store technology which is gaining momentum in data protection and could possibly be one of the necessary features in any future data protection application. We described in length the potential use and suitability of this technology in the protection of remote offices. We concluded the chapter by discussing how the integration and usage of disk by data protection applications is gradually changing the pricing models commonly used today by most data protection application vendors. The machine or CPU based model is rapidly becoming inadequate in the world of server consolidation and virtual computing. A capacity-based model and its advantages and disadvantaged were discussed to give the reader an idea of the potential changes that may lay ahead.

Some of the biggest impact to requirements of backup and restore has come in the guise of compliance. Managing large amounts of data now has large business impacts in many facets of organizations and corporations. Many are still struggling with what the requirements are and how to manage them. This area will most likely grow significantly in the next few years. Technology introduced will help to meet the growing compliance needs and make it easier to define and implement policies around them.

We have taken time to identify areas of focus and discuss the remarkable changes that the area of data protection has under gone. It is no longer about passively managing backups and restores. The paradigms have shifted significantly to require a breakthrough into a completely new area and focus: that of active management. Data protection is now an ingrained, a required and a critical part of server operations.

Glossary

Access Control	The granting or withholding of a service or access to a resource to a user or service based on identity.
ACL	Access control list – a table that defines what access rights each user has to a particular system object, such as a file directory or individual file. Each of these objects has a security attribute that identifies its access control list. The list contains an entry for each system user with access privileges
Adware	Any software application in which advertising banners are displayed while the program is running.
Algorithm	A derivation of the name of the Arab mathematician Al-Khawarizmi, meaning a well-defined step-by-step process for solving a problem.
Audit trails	A log that is used to track computer activity. This is generally used to track access.
Backup	The process of making a copy of the data from a system that can be preserved in case of equipment failure or other catastrophe.
Backup window	The time or 'window of opportunity' given to the backup process in which to execute either scheduled or user directed backups.

Digital Data Integrity   David Little, Skip Farmer and Oussama El-Hilali
© 2007 Symantec Corporation. All rights reserved

BMR	Bare Metal Restore – The process by which the entire system is recovered onto hardware that does not yet have an operating system installed.
Business impact analysis	A methodology that helps to identify the impact of losing access to a particular system or application to your organization.
Catalogue	A directory of information about data sets, files, or a database that usually describes where the data set, file or database entity is located and may also include other information, such as the type of device on which each data set or file is stored.
CPU	Central processing unit – an older term for processor and microprocessor which is the central unit in a computer containing the logic circuitry that performs the instructions of a computer's programs.
Client – backup	Any system that contains data that will be preserved by a backup application.
Cluster	A group of servers and other resources that act like a single system and enable high availability and, in some cases, load balancing and parallel processing.
Cluster node	A single system that is a member of a cluster.
Compliance	A state of being in accordance with established guidelines, specifications, or legislation.
CDP	Continuous data protection – a storage system in which all the data in an enterprise is backed up whenever any change is made.
COW	Copy on write – a snapshot that consists of a list of blocks whose contents have changed as snapshot initiation and a private data area containing the blocks' contents prior to the change.
Cumulative incremental backup	A type of backup that will collect all files that have changed, as the last successful full backup. All files are backed up if no prior backup has been done.
DAS	Direct Attach Storage.

Data Center	A centralized location, either physical or virtual, for the storage, management, and control of the data and information for a particular body organization or business.
DCL	Data change log – a log that stores updates to the primary as the associated snapshot volume was created. It is used for mirror fast resynchronization.
Decryption	The process of converting encrypted data back into its original form, so it can be understood.
DHCP	Dynamic host configuration protocol – TCP/IP protocol that automatically assigns temporary IP addresses to hosts when they connect to the network.
Differential incremental backup	A backup that copies the files that have changed as the last successful backup of any kind. All files are backed up if no prior backup has been done.
Disaster recovery	The act or process of recovering data from backups after a disk crash or other catastrophe.
Disk staging	An automated process for using disk as a cache during the backup process that will move the backed up data from the disk to a secondary location.
Encryption	The process of converting data into a form, called a ciphertext, that cannot be easily understood by unauthorized people.
Expunge	To remove, erase, or completely delete data or references to the data. Most common usage of this is the removal of legal records when requested by the court.
Fast resynchronization	The ability to resync a mirror based on a log of the changes made while the mirror is split.
File segment	A variable portion of a file as determined by the single instance store application.
Firewall	A set of related programs which can be located at a network gateway server that protects the resources of a private network from users to other networks.

FOIA	Freedom of information act – law requiring government agencies to provide information to the public when requested in writing. B24.
Full backup	This is a backup that copies to a storage unit, all files and directories that are beneath a specified directory.
High availability	The ability of a system to perform its function continuously and without interruption for a significantly longer period of time than the reliabilities of its individual components would suggest.
HIPAA	Health insurance portability and accountability act – an act that deals with protecting health insurance coverage for people who lose or change jobs and includes an administrative simplification section which deals with the standardization of healthcare-related information systems.
HSM	Hierarchical storage management – the application or technology that manages the movement of data between different types of storage based on time or access.
Image	The collection of data that NetBackup saves for an individual client during each backup or archive. The image contains all the files, directories, and catalogue information associated with the backup or archive.
Incremental backup	A backup that copies changed files. See cumulative incremental and differential incremental backup.
JBOD	Just a bunch of disk – disks that share an enclosure but had no intelligent interface.
Key management	This generally refers to the process of managing the keys used in encryption so that the data can be retrieved if needed.
LAN	Local area network – a group of computers and associated devices that share a common communications line or wireless link.
LDAP	Lightweight Directory Access Protocol.
MAID	Massive array of idle disk – a storage technology that employs a large group of disk

	drives in which only those drives in active use are spinning at any given time.
Master server	The backup server that provides administration and control for backups and restores for all clients and servers within a backup domain.
Media	This is the physical magnetic tapes, optical disks, or magnetic disks where data are stored.
Media server	The backup server that actually moves the data from the client system to the backup storage whether it be tape or disk.
Mirror copy	An identical copy of the original data that is usually kept in sync.
NAS	Network Attached Storage.
Operating system	The program that, after being initially loaded into the computer by a boot program, manages all the other programs in a computer.
Patch	The code that is provided to correct a problem or problems in the original program or application.
Primary copy	The main or original copy of data.
Provisioning	The automation of all the steps required to manage (setup, amend, and revoke) user or system access entitlements or data relative to services or systems.
PST	Personal storage – a personal folder file in Microsoft Outlook.
RAID	Redundant array of independent (or inexpensive) disks - a way of storing the same data in different places on multiple hard disks.
Remote office	Generally used to refer to data that exists in a remote location that does not have high-speed network access to the central data center.
Replication	The process of making a copy of something.
Restore	A process that involves copying backup files from secondary storage to hard disk in order to return data to its original condition if files have become damaged, or to copy or move data to a new location.

SAN	Storage Area Network.
Sarbanes-Oxley	Legislation enacted that defines which business records, including electronic records and electronic messages, are to be stored and for how long.
Scalar multiplication	The multiplication of a vector by a constant.
Schedules	Controls when backups can occur in addition to other aspects of the backup, such as: the type of backup (full, incremental) and how long the backup application retains the image.
Secondary storage	All data storage that is not currently in a computer's primary storage or memory.
SATA	Serial advanced technology attachment – a new standard for connecting hard drives into computer systems based on serial signalling technology.
SLA	Service level agreement – a contract that specifies, usually in measurable terms, what services are going to be provided to a customer or user.
SNMP	Simple network management protocol – a mature standard that was developed in response to the need to manage peer network elements supporting the extraction of information from the managed element (SNMP GET), the setting of parameters that control the managed element (SNMP SET) and the processing of event signals emitted by the managed element SNMP TRAP).
SIS	Single instance store – the process of identifying duplicate data patterns and linking them together.
Snapshot	A persistent copy of a data volume or set of pointers to the data volume that is frozen in time.
Spyware	Adware that includes code that tracks a user's personal information and passes it on to third parties, without the user's authorization or knowledge. Any software that employs a user's Internet connection in the background without their knowledge or explicit permission.

SRM	Storage resource management – a tool that reports on storage utilization across the enterprise.
Synthetic backup	Any backup that is created without actually accessing the original data source.
Synthetic full	A full backup that is created without actually accessing the original data volume.
VTL	Virtual tape library – an archival storage technology that makes it possible to save data as if it were being stored on tape although it may actually be stored on hard disk.
Virtualization	The creation of a virtual version of something, such as an operating system, a server, a storage device or network resources.
Virus	A program or programming code that replicates by being copied or initiating its copying to another program, computer boot sector or document.
Vulnerability (security)	A weakness in an operating system, system security procedures, internal controls, or implementation that could be exploited.
WAFS	Wide area file service – a storage technology that makes it possible to access a remote data center as though it were local.
WAN	Wide area network – a communications network that is geographically dispersed and that includes telecommunications links.

Appendix A

**Backing Up VMware With
Veritas NetBackup™**
Best Practices
NetBackup 6.0 and
VMware ESX Server 2.x and 3.x

Digital Data Integrity   David Little, Skip Farmer and Oussama El-Hilali

White Paper: Enterprise Solutions

Backing Up VMware With Veritas NetBackup™ Best Practices
NetBackup 6.0 and
VMware ESX Server 2.x and 3.x

Contents

EXECUTIVE SUMMARY

Server virtualization is quickly becoming a standard technology in many data centers today. While it can significantly augment computer utilization, this technology introduces new issues related to protecting virtual environments. Data created and utilized in virtual machines is no less important than data located in a single physical machine. This paper describes several approaches that can be used to back up VMware ESX Server 2.x and its underlying components using Net-Backup software. This paper also discusses the relative advantages and disadvantages of each method. Unless otherwise specified, each of these techniques is fully supported by both Symantec and VMware.

VMWARE ESX SERVER OVERVIEW

Before discussing specific backup technologies, let's discuss the structure of a VMware instance (see Figure 1) and the data protection issues that exist because of this structure. The VMware kernel runs within a Red Hat Linux instance that is optimized specifically for the VMkernel. This optimized Red Hat instance runs two file systems: ext3 and the VMware ESX Server file system (VMware Virtual Machine File System, or VMFS). Smaller configuration files (e.g., ".vmx" files) are stored on the ext3 file system. Much larger virtual disk images are stored on VMFS. VMFS has been specifically optimized for the large vmdk files that commonly exist in VMware environments.

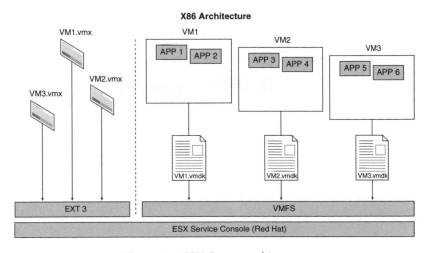

Figure 1.   ESX Server architecture

BACKING UP VMware ESX SERVER 2.X

While VMware can provide enhanced server utilization and flexibility, it also introduces unique backup issues:

- Do you back up the individual virtual machines as clients or do you back up the underlying vmdk files on which the virtual machines are based?
- While one virtual machine is being backed up, what is the impact on additional virtual machines hosted on the same physical server?
- What are the relative advantages of each of these backup techniques?

There are many ways that NetBackup can be configured to safely protect VMware environments. Taking into account these issues, we discuss these methods in detail.

Method 1: Back up the guest OS as an NetBackup client

Backing up standard client data

This technique is probably the easiest to implement and the most straightforward way of backing up an operating system running within VMware. A standard NetBackup client is simply installed on the guest OS (see Figure 2). The guest OS backup is then scheduled and performed as you would with any other NetBackup client. File restores are the same as they would be for any standard client.

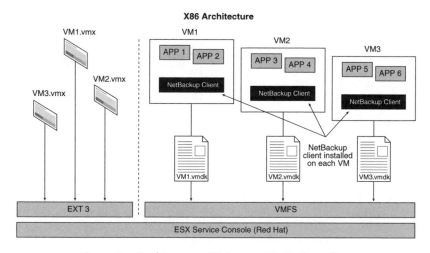

Figure 2. Backing up a VM as an NetBackup client

Backing up database data (or any complex data structure)

The basic principles that apply for standard client data also apply for database data. If the database hosted on the guest OS is supported by NetBackup, the appropriate database agent can be installed and configured, and the database safely backed up. If the database or complex data structure is unsupported, then NetBackup pre- and post-processing scripts can be utilized to ensure that the data structure is in a consistent, stable state before the backup process begins.

Considerations

While this technique is straightforward to implement, it provides advantages and disadvantages. On the positive side, this backup methodology is typically the same as other backup clients in the environment. Backups and restores occur just as if the guest OS was hosted on a physical, not virtual, machine. As long as the guest OS is supported by NetBackup and VMware, it can be backed up safely. The disadvantages with this method are several. Entire system restores can be problematic with this methodology. Backup processing load needs to be taken into account as well. Because backup activities are CPU- and I/O-intensive operations, it is recommended that each individual guest OS is backed up serially or one at a time to minimize the impact that backups have on other guest OSs hosted on the same physical machine.

NetBackup configuration

When backing up the virtual machine as a client, configuring NetBackup is straightforward. The NetBackup client policy is simply configured as you would any standard NetBackup client, depending on which operating system the guest OS is running. The key to minimizing the impact on other virtual machines hosted on the same physical server is scheduling for each of the virtual machines. Keep in mind that virtual machines hosted on a single physical server share the same finite resources of that server. Because backup activities tend to be I/O- and resource-intensive, best practices dictate that each guest OS hosted on a single physical server be backed up serially. In this way the impact of backup operations on the physical host can be minimized and

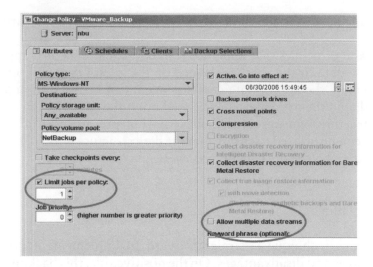

Figure 3. NetBackup policy configuration

the impact on other virtual machines on this same host, mitigated. To facilitate this, proper NetBackup configuration is required as follows:

On the NetBackup server, configure these policy attributes (see Figure 3):

- Limit jobs per policy—select this attribute and set it to 1
- Allow multiple data streams—deselect this attribute

The effect of this configuration setting is that only one virtual machine client per NetBackup policy will be backed up at a time. All backups will be serialized, and the impact of backup processing on the remaining guest OSs will be minimized.

Method 2: Back up the VMware disk (vmdk) files

Each VMware guest OS has at least one physical disk file associated with it. VMware places these files on top of VMFS and adds a "vmdk" extension to each file. These files can be backed up as standard files. In this configuration, a NetBackup client is installed on the VMware service console, which is running Red Hat Linux (see Figure 4).

Restoration of a virtual machine would simply require the restoration of the individual vmdk file(s) associated with that virtual machine.

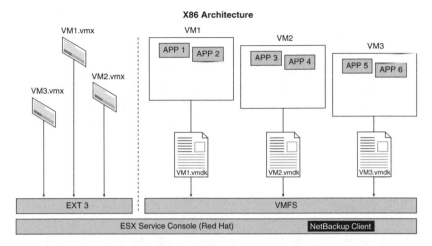

Figure 4. NetBackup client installed on a service console

Care needs to be taken when backing up these files to make sure they are backed up in a consistent manner, assuring that restored files are not corrupt. Before backing up the vmdk files, all I/O operations to these files must be halted. There are three ways to accomplish this:

1. Power off or suspend the virtual machine.   I/O to the vmdk files must be halted, and powering off the guest OS is probably the simplest way of accomplishing that. Unfortunately, this technique dictates that the virtual machine be unavailable for the duration of the backup.

2. Create a snapshot of the virtual machine. This technique uses snapshotting capabilities built into ESX Server (not to be confused with a volume or file system snapshot). While the virtual machine is operating, ESX Server can stop writes to the vmdk files by invoking the ESX Server snapshotting capability. All writes to the vmdk files are halted, and new writes are captured by a redo log (a bitmap record of changes to the original disk). The vmdk files are now static, and backups can proceed. During backup operations, the guest OS is always available. Once the backups are completed, redo information is applied to the vmdk files and the redo logs are deleted. This technique must be used with care as the ESX Servre snapshot technology leaves the backed-up OS and applications that are based on vmdk files in what is considered a "crash consistent" form. This means that there is no guarantee that a restore from this

condition will be usable. Symantec does not recommend using this technique.

3. Power off the virtual machine, then create a snapshot. In this technique the virtual machine is powered off, a snapshot is quickly created, the machine is powered on, and then backups proceed. The machine is now available to users. This method provides a compromise backup methodology that limits the downtime of the virtual machine while ensuring that the virtual machine backup is consistent and recoverable.

Considerations

Backing up the vmdk files associated with the virtual machines is straightforward and provides certain advantages and limitations. These include:

- If each virtual machine located on a single physical machine is backed up serially, it is recommended that you limit the number of file systems that are backed up in parallel, to minimize the I/O impact that backup operations might have on running virtual machines.
- All vmdk files associated with a specific virtual machine must be backed up at the same time. If the vmdk files are not backed up at the same point in time, data loss or corruption can occur. At this point it is recommended that the VMware configuration files or .vmx files be backed up as well. These files are small and typically don't change often, but it is recommended that they be backed up every time the guest OS vmdk files are backed up.

Backing up the ESX service console

VMware creates the ESX service console structure on top of a modified Red Hat Linux operating system. This structure includes files that exist on two file systems: ext3 and VMFS (see Fig 1). Because the service console is an OS supported as a NetBackup client, backing up the service console itself is a simple matter of installing a standard NetBackup Red Hat Linux client on the console and backing it up using a standard NetBackup policy. (Note: This is essentially the same configuration as backing up the vmdk files. Both the service console files and VMware associated files are backed up with this client.)

Considerations

The service console itself (excluding .vmx and .vmdk files) docs not typically change often, so backing up the service console OS does not need to occur very frequently. However, in the event of a system failure, restore operations would be facilitated if a current backup of the service console was available.

As a point of clarification, the backup of the service console described in this section would not typically include the .vmx and .vmdk files associated with a guest OS. These files would be protected separately as mentioned in Method 2 (Back up the VMware disk (vmdk) files).

Solution comparison chart

The following chart provides a comparison of the two backup technologies described in this document.

NetBackup and VMware Backup Solution Comparison

Functionality	Guest OS as NetBackup Client	Backup Virtual Disk Files
File-level backups	Yes All client files or a subset of the client files may be backed up.	No The entire vmdk file must be backed up. Partial are vmdk file backups not possible.
Individual file restore	Yes One or any number of files may be restored. True Image Recovery may be used as well.	No No single file or partial restores are possible. The entire vmdk file must be restored.
Incremental backups are supported	Yes	No
Can utilize synthetic backup technology	Yes	No
Point-in-time backup is possible	Yes	Yes Because this backup is at the vmdk file level, all backups are effectively point-in-time backups. The amount of data backed up can be significantly larger, as every backup requires entire vmdk file(s) to be backed up.

(*Continued*)

Functionality	Guest OS as NetBackup Client	Backup Virtual Disk Files
Backup load issues	**Yes** Full and incremental backups can create a lot of I/O, which in turn can use significant resources. Staggered backups can mitigate this load.	**Limited** The load on the system can be isolated to I/O activity on a specific vmdk file.
Simplified DR	**No** DR operations can require significant planning and expertise.	**Yes** Disaster recovery can be as simple as restoring the vmx and vdmk files.
Must restore entire VMware disk file (.vmdk) to restore a single file	**No** Granular file level restores are possible.	**Yes** Single file restores are not possible. The entire vmdk file must be restored to restore a single file.
backing up the service console is recommended	**Yes** No matter what backup technique is used, backing up the service console is always recommended.	**Yes** No matter what backup technique is used, backing up the service console is always recommended.
Guest OS must be shut down or suspended during backup	**No** The guest OS can be 100 prcent available during backup operations. Open files are backed up using standard VSS or VSP technologies built into NetBackup.	**Yes** It is recommended that the guest OS be shut down to insure that a consistent backup is achieved.

RUNNING A NETBACKUP MEDIA SERVER IN A VMware ENVIRONMENT

At first glance this configuration seems to make a lot of sense. Positioning the NetBackup Media Server close to the data requiring backup seems logical enough, but there are significant issues with this configuration. The main point to keep in mind is that a NetBackup Media Server (or any backup server) requires a significant amount of system resources during backup operations. Because of the I/O-intensive nature of backup, adding a Media Server to a virtual environment that is most likely busy to begin with does not make sense from a resource management perspective. VMware is designed to make use of all unused system resources on a given server or host. Adding backup activity to a host that is already busy supporting multiple VMs simply does not make any sense.

There are two commonly referenced configurations for placement of the NetBackup Media Server. The first is placing the Media Server within a virtual machine, and the second is placing the Media Server on the ESX service console.

Media Server in a virtual machine

This technique is seemingly attractive in that the NetBackup Media Server is as physically close to the virtual machines as possible. But there are two issues that must be considered.

First, most servers hosting numerous virtual machines typically do not have enough computing and memory resources to support a Media Server running in an additional virtual machine. The impact of backup operations on the other guest OSs hosted on this server can be enormous.

Second, VMware does not currently support fibre-attached tape drives. This significantly limits flexibility when tape drives and libraries are incorporated in the backup environment.

Media Server in the ESX service console

This configuration is not supported by VMware. Because VMware does not recommend or support this configuration, NetBackup does not support it either.

Keep in mind that the service console is running in an optimized version of Red Hat Linux. This modified version of Red Hat Linux was never intended to support applications such as backup servers.

Reference materials

Title: "VMware ESX Server Backup Planning"
Desc: A VMware created doc that covers backing up VMware ESX Server 2.x
URL: www.vmware.com/pdf/ESXBackup.pdf

Title: "Consolidated Backup in VMware Infrastructure 3"
Desc: Covers the VMware Consolidated Backup functionality that was introduced
    with ESX Server 3
URL: www.vmware.com/pdf/vi3_consolidated_backup.pdf

Title: Backup Software Compatibility for ESX Server 2.x
Desc: VMware's compatibility matrix
URL: www.vmware.com/pdf/esx_backup_guide.pdf

Title: Symantec Support Site
Desc: VMware related support information, including NetBackup client compat-
    ibility
URL: http://support.veritas.com

ABOUT SYMANTEC

Symantec is the world leader in providing solutions to help individuals
and enterprises assure the security, availability, and integrity of their
information. Headquartered in Cupertino, Calif., Symantec has operations in more than 40 countries. More information is available at
www.symantec.com.

   For specific country offices and contact numbers, please visit our Web
site. For product information in the U.S., call toll-free 1 (800) 745 6054.

Symantec Corporation
World Headquarters
20330 Stevens Creek Boulevard
Cupertino, CA 95014 USA
+1 (408) 517 8000
1 (800) 721 3934
www.symantec.com

Appendix B

Secure Optimized Data Protection for Remote Offices
An Overview of Veritas NetBackup PureDisk™ Remote Office Edition

Wim De Wispelaere
Senior Manager, Product Management

Digital Data Integrity David Little, Skip Farmer and Oussama El-Hilali

White Paper: Enterprise Solution

Secure Optimized Data Protection for Remote Offices White Paper
An Overview of Veritas NetBackup PureDisk™ Remote Office Edition

Contents

DATA MANAGEMENT LANDSCAPE

Many organizations are struggling with the massive changes in data storage requirements that have transpired over the last decade. The almost exponential growth of business-critical data from email, e-commerce, and electronic systems shows no sign of decreasing. With relatively new data types such as voice and video now in use, enterprise storage administrators will soon have to manage petabytes of data.

According to a recent study by the School of Information Management and Systems at the University of California at Berkeley, the world produces between 1 and 2 exabytes of unique information per year, which is roughly 250 megabytes for every man, woman, and child on earth. An exabyte is a billion gigabytes and printed documents of all kinds comprise only .003 percent of the total. Magnetic storage is by far the largest medium for storing information and is the most rapidly growing, with shipped hard-drive capacity doubling every year. Magnetic storage is becoming the universal medium for information storage.

The annual growth rate of corporate reference data is estimated to be 60 percent. Traditional "weekly full, daily incremental" backup approaches are ill suited to cope with this situation: companies should work 60 percent more efficiently to prevent costs from increasing, which is not very likely to happen with the current systems. As the business world has moved to a 24-hour, 7-day-a-week working cycle, the notion of overnight "downtime" for maintenance and backup is less feasible. Symantec has developed software to help companies manage data integrity more efficiently and meet today's standards for data protection.

The rise in the volume of data has seen a corresponding tightening of corporate governance and legal procedures surrounding the retention and availability of data. According to one large storage vendor, there are over 4,000 major regulations that apply to information-keeping worldwide. The United States has the most, with federal statutes such as the Health Insurance Portability and Accountability Act (HIPAA) that covers medical records, and the Food and Drug Administration Section 21 rules that carry heavy fines for noncompliance with data-retention rules. The most commonly quoted U.S. regulation is the Sarbanes-Oxley Act of 2002, which was brought into force after Andersen employees shredded important documents in the wake of the Enron scandal. The Securities and Exchange Commission (SEC) also has extensive rules governing data retention, with heavy fines and even jail sentences for executives in cases of noncompliance.

The U.S. National Archives and Records Administration (NARA) and the United Kingdom's Public Records Office are two examples of bodies whose entire business is focused on ensuring records are maintained correctly and effectively. In fact, most developed countries have similar governmental structures to ensure that data is archived and released within a legal framework. These guidelines, in turn, must be adhered to by other governmental agencies such as the Social Security Administration and the Internal Revenue Service, as well as banks and building societies.

In Europe, draft European Union (EU) legislation being formally ratified now will force telecom companies and Internet service providers (ISPs) to retain information on their customers' logs of phone calls or e-mail and Internet connections beyond the one- or two-month period the information is normally held for billing purposes. The period could be up to a year; this change is intended to assist police and fraud investigations.

DATA AND RISK EVOLUTION

Conventional data management and protection techniques have not kept pace with the increasingly complex nature of today's data processing and topology. Many IT organizations still use the "weekly full, daily incremental" backup technique employed since the 1950s. In the past half century, topologies have evolved from centralized homogeneous platforms to heterogeneous networks of distributed and mobile systems with multiple storage tiers; annual data growth has increased from 20–35 percent to 80–100 percent; retention periods have increased from weeks to decades; and usage patterns have evolved from transaction to transaction and reference. Studies indicate approximately 60 to 80 percent of this growth is fuelled by reference data. Reference data describes information with access requirements measured in seconds to minutes while transaction data describes information with access requirements in milliseconds.

Traditional data-protection and management approaches fail. Ernst & Young's Fabric of Risk study determined that approximately 36 percent of the executives from the top 1,000 publicly traded companies believed their companies would cease operations due to inadequate protection, while 59 percent placed their risk as moderate to high. The increased dependence upon networked and mobile data, combined with theft and vulnerability to viruses, exacerbates this risk. New technologies such as storage area networks (SANs) aid physical storage management, but affordable data-protection and management solutions have been elusive.

PRESENT-DAY ISSUES WITH CLASSIC TECHNOLOGY

The globalization of businesses and cultural changes such as home and mobile working have made the backup process more complex. Companies operating from different local offices must distribute parts of their IT infrastructure over these remote sites out of necessity. Local documents, emails, presentations, and so forth are kept on local file servers primarily to improve network performance and to allow rapid recovery in the event of data loss (Figure 1).

Many enterprises organize the backups of their remote sites locally, rather than sending data over a wide area network (WAN) to a central backup server. Local backup, however, isn't without its hazards:

- It can hinder central control and monitoring and hence introduces the potential for errors.
- The people performing remote backups may not follow central procedures and security policies exactly as specified. Are tapes systematically transferred to a vault and stored in a secure environment that protects them against changes in temperature and humidity? Is the backup executed every day, as stated in the guidelines?

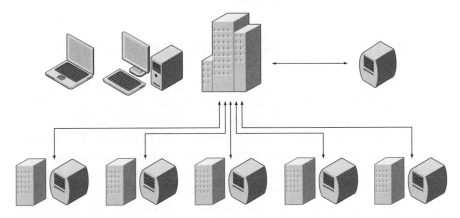

Figure 1.   The volume of files on remote office servers continues to grow rapidly.

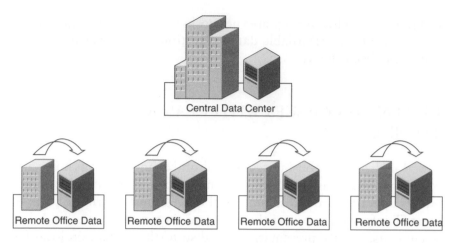

Figure 2.   Remote office backup can be cumbersome and problematic.

- Backup and recovery operations can fail due to a lack of skilled resources, lack of media, or other technical problems not easily solved remotely.

As many studies and onsite audits have pointed out, the backup of data at a remote site is often executed by non-IT personnel. Sometimes the wrong tape is inserted and there is no one else present to check it. The response to system errors is often incorrect and is not reported to a central IT authority. Worse, when the tape loading process fails, no backup is executed. As the remote employee is often not qualified to verify whether the backup has been successful or not, no one really knows if the tapes contain the correct data.

LIMITATIONS OF THE CURRENT WAN SOLUTION

The number and size of files stored on local file servers are increasing dramatically. Consequently, performing a full backup of remote data over a WAN connection requires a considerable amount of bandwidth and can be prohibitively expensive. As the number of remote sites increases, this problem becomes an even bigger issue, creating a bottleneck between the remote site and the data center's backup server. As data volumes grow and working patterns edge towards 24 × 7 operation, the overnight backup window may not even be a feasible option. This leaves many sites with only one option: to

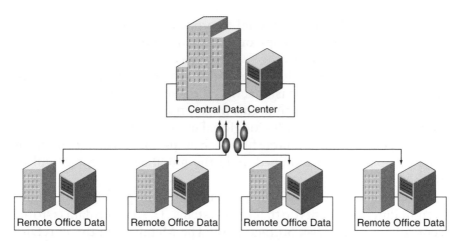

Figure 3. Corporate wide-area networks (WANs) can not handle traditional backups of remote offices.

organize all backups locally with backup tapes stored at the site or sent to an offsite facility.

NETBACKUP PUREDISK BUSINESS PROPOSITION

The massive growth in data generation and retention periods, combined with legislative requirements, requires a fundamental change in backup procedures. This is especially true with remote offices and mobile users who have often been outside of the scope of a centralized IT infrastructure.

Symantec is a pioneer in the field of content routing, a technology that provides a long-term framework to address both the rapidly growing volume of data and the wider information lifecycle issues. Veritas NetBackup PureDisk is a software solution for the protection of file data on clients anywhere on the network to any type of disk-based storage pool.

With its unique fingerprint technology called global unique file identification, NetBackup PureDisk technology distinguishes unique files from redundant copies across the enterprise. Enormous savings in storage capacity and network traffic are achieved by not transmitting and storing redundant data.

For example, a backup across three remote offices of the same two-megabyte Word file on three file servers would result in six megabytes of capacity used in conventional approaches. NetBackup PureDisk, however, stores a single copy only and consequently needs no more than two megabytes of storage capacity—a 66 percent savings. Comparable results are seen for throughput: The backup would be completed about 66 percent faster across the three remote sites.

Through file segmentation, this fingerprint technology can even be applied on small parts of files for global unique segment identification. This is typically done for large files such as .pst files. When such a file is backed up and then modified, only the modified segment(s) will be backed up.

Global unique file and segment identification is performed by lightweight PureDisk agents on the client systems. The backup data is stored centrally in a PureDisk Storage Pool. The Metabase in the PureDisk Storage Pool will store the file metadata in a scalable, distributed database. The file content segments are stored in one or multiple Content Routers in the Storage Pool. Because metadata and content data are stored separately, all source file versions can be restored, while only the globally unique file segments are stored.

In a typical system, about 13 percent of the files are modified each day and must be backed up. Using PureDisk global unique file and file

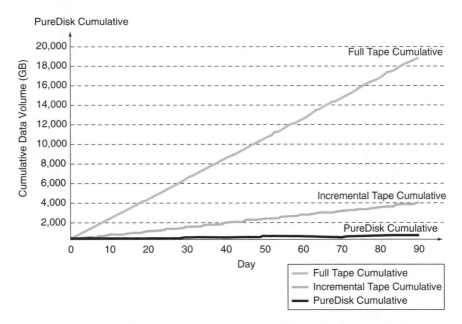

Figure 4. Growth of backup data volume using different backup methods.

segment identification, our studies show the number of actual bytes changed is 1-2 percent. For example, say a company has one terabyte of data. If it performs four incremental and one full backup each week over a period of 16 weeks, these 80 backups would require more than 16 terabytes of storage. PureDisk will store an optimized initial backup and then 16 weeks of incremental backups. After 16 weeks, PureDisk will typically need 1.5 terabytes to store all versions of all files within this retention time. With 80 terabytes of source data protected in 80 backups of one terabyte each, PureDisk has reduced the backup data volume by more than 50 times to 1.5 terabytes.

PureDisk agents installed on the clients perform the global unique file and segment identification locally and only send incremental data over the network. This can reduce the backup bandwidth by a factor of 50. Figure 5 illustrates the bandwidth savings of NetBackup PureDisk versus traditional methods. Because only new and unique file segments are transmitted, present-day WANs such as Internet VPN have sufficient bandwidth for the transfer of remote data to the data center.

NetBackup PureDisk reduces the maintenance cost related to backing up data at remote offices. Typically, such maintenance involves

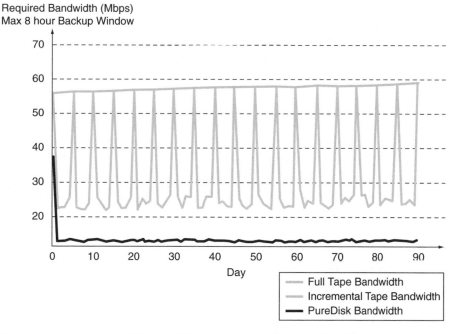

Figure 5.   Comparison of bandwidth required to replicate 1 TB of data over WAN within 8 hours.

system maintenance, tape management, and the shipment of media to a vault site. NetBackup PureDisk performs automated disk-based back-ups and can be remotely operated using the Web-based user interface. This means NetBackup PureDisk does not require anybody to operate the system or handle tapes at the remote site. Eliminating manual backup system handling will result in a higher backup-and-restore success rate.

Because PureDisk backup data is replicated automatically to the central office and/or a recovery site, there is no need to ship physical media. This results in additional cost savings.

NetBackup PureDisk performs a data reduction of 50 times on the backup data. As a result, smaller amounts of data need to be managed, which requires fewer storage operators. Because of the data reduction, backups will be completed faster. In combination with the parallel processing capability of NetBackup PureDisk, this results in a signifi-cant reduction in overall backup time.

NETBACKUP PUREDISK IMPLEMENTATION

A typical environment protected by NetBackup PureDisk has a Pure-Disk Storage Pool at each site for local backups and a central Storage Pool for replication of the local backup data. However, PureDisk technology can also protect smaller environments without the need for a local Storage Pool. Figure 5 displays both of these deployment scenarios.

The agents on all systems send the new, unique file segments to a local PureDisk Storage Pool, which verifies the uniqueness of files across all local agents. It then replicates only the unique file segments to the main PureDisk Storage Pool. During this replication, the unique-ness of the file segments is checked across all connected agents and Storage Pools. This minimizes WAN bandwidth requirements and allows long-term scalability because of the reduced Storage Pool capa-city requirements.

A PureDisk Storage Pool on the remote site optimizes the data over all the office's clients by identifying unique contents and stores backup data locally. This shortens backup-and-restore tasks, and enables synchronization with central or distributed locations independent of the local backups.

While backups are performed over the LAN during business hours, synchronization to the central data center can be performed overnight,

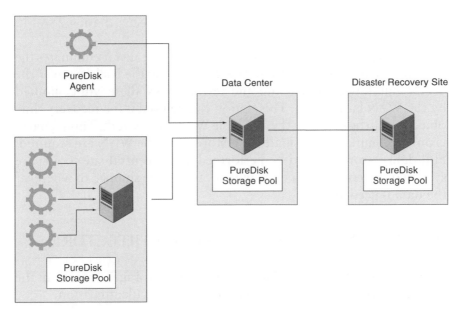

Figure 6. Illustrates how agents can backup data to local or remote storage pools and how these storage pools can efficiently replicate to other remote storage pools.

when the WAN has less activity. This reduces WAN traffic back and forth between the data center and remote sites.

Using PureDisk replication, there are always two full remote-site backups—one local, and one on a central server—every day. As an option, backup data can be replicated again to a disaster recovery site, adding another level of data protection.

In a replication policy, data filters can be applied. Using filters, only selected files or file types will be replicated to the central Storage Pool.

Data retention in the PureDisk environment is managed by data removal policies per Storage Pool. In the remote office, a data retention policy can be defined to retain all versions of files during a short period of time, typically a few weeks. On the central Storage Pool, a data retention policy for the replicated data can be configured to keep all versions of all files for a longer period of time, typically several months. Using the PureDisk data retention features, the local PureDisk Storage Pool can be configured as a local backup cache allowing fast local restores, while the central Storage Pool can be configured as a medium-term data archive.

Smaller remote offices with smaller amounts of data or fewer clients to protect can opt not to have the local PureDisk Storage Pool at the remote site. Instead, they can back up files instantly over the WAN to the central Storage Pool. All files, however, are secured using 256-bit, key agent-based encryption.

This solution is less expensive to implement because it eliminates the need for a device at the remote site. Performing the backups over the WAN will slow down backup-and-restore tasks; these will be limited by the WAN bandwidth. This option can be used if the recovery time objective can be met with the available WAN bandwidth. This scenario is only recommended if single file restores over the WAN are required. Full data restores should be performed at the central site to a spare system or to removable media, which can then be shipped to wherever the data is needed.

PUREDISK TECHNOLOGY AND ARCHITECTURE

Symantec uses the term Global Single Instance Storage to represent the combination of Global Unique File and Segment Identification.

Global unique file identification

PureDisk employs a distinctive fingerprint technology to distinguish unique files from redundant copies. The fingerprint is derived from the total binary contents of the file. The result is that files with the same content will have the same fingerprint, even when the files have different names, locations, attributes, creation or modification dates, and security. Only one copy of a file with a certain fingerprint will effectively be sent to the PureDisk Storage Pool. Other copies of the same file will be identified because they will generate the same fingerprint. Even when the file name and path of the copied file is different or when the files are stored on geographically distant systems, NetBackup PureDisk will identify these files to be exact copies based on their identical fingerprints. By using globally unique fingerprints, the disk space required for the PureDisk Storage Pool grows at a much slower rate than by using traditional backup methods.

Global unique segment identification

NetBackup PureDisk also divides larger files into smaller segments with a configurable segment size. By cutting a large file into small chunks of data, the system will only back up the parts that contain

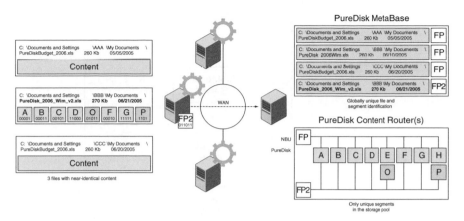

Figure 7.   Global Unique File and Segment Identification.

new or modified data, while the majority of the file remains unchanged and doesn't need to be backed up again.

Separated storage of metadata and content data

Agents send the backup data to a PureDisk Storage Pool. In a PureDisk Storage Pool the file properties are stored in a Metabase, and the file contents are stored in Content Router(s).

For all files backed up, including file versions and deletions, the file metadata is stored in the Metabase. The Metabase is a scalable, fully searchable, distributed database for metadata. The Metabase contains the file metadata such as name, path attributes, and security settings, together with the file fingerprint.

The unique file contents are stored in one or more Content Router(s). The first characters of the segment fingerprint determine the position where the file segment is stored in the Content Router(s) according to the content routing table.

With all metadata of all file versions in the Metabase, you can restore any version of any file to any client, while in the Content Router(s) only the necessary unique file segments have been stored.

Metabase

The PureDisk Metabase stores all metadata of files and previous versions, together with their fingerprints. The front end of the PureDisk

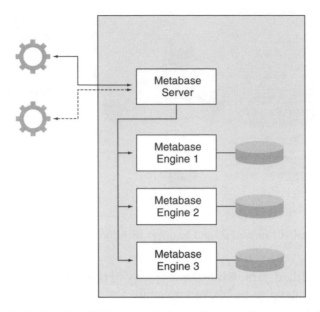

Figure 8.   NetBackup PureDisk uses a 2-tier architecture for storage of metadata.

Metabase is called the PureDisk Server. The PureDisk Metabase Server delivers a fully searchable view across all the stored metadata. The actual metadata is stored in one or more PureDisk Metabase Engines (see Figure 8). The Metabase Engines are relational database systems. Through its two-layer architecture, the Metabase can be scaled in size and performance by simply adding Metabase engines.

Users can query the Metabase to locate a particular file or set of files. The Metabase search tools allow the user to specify many different file-property and time-related search criteria. Apart from its obvious assignment as a keeper of backup metadata, the Metabase is also a valuable tool for data assessment. For example, a user could query the Metabase to identify all data older than three months across all of its remote file servers, or all data that was not accessed in the last month. Complex questions can be answered by combining queries and post-processing the results. For example:

- How much storage space do I currently need to store all my documents or all my presentations?
- How much duplicate information is currently stored?
- How many versions of this document do we have?
- Which systems use this infected dll?

Content Router

A Content Router is responsible for storing and restoring the file segment data from and to the PureDisk Agents. A Content Router can also redirect file content data to another Content Router.

Each Content Router is connected to a storage device of any type (direct attached storage [DAS], network attached storage [NAS], SAN). When receiving content-enabled data for which the current Content Router is the final destination, it will effectively store this data on its connected storage device.

Each Content Router is responsible for a specific subrange of the full PureDisk content address range. The address of a data object is derived from the unique fingerprint of this object. The distribution of the content addresses across the available Content Router(s) is determined in the content routing table.

The number of Content Routers in a PureDisk environment depends on the amount of data to be dealt with and the data traffic. Scalability is achieved by simply adding more Content Routers (with attached storage) when the need arises without having to interrupt or modify the backup or restore policies. This provides online scaling beyond petabytes of data within a single Storage Pool. The expansion of the total volume of storage can happen while the PureDisk system remains online. With the addition of new Content Routers with storage, the content

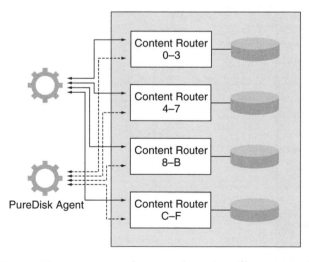

Figure 9.   Content Routers store and restore the unique file segment data through the PureDisk Agents.

routing tables will be updated to reflect the new configuration. From that moment, the Content Routers will redistribute the existing content data according to the new storage organization. This happens in the background while the system remains available for backups and restores.

Web-based user interface

All PureDisk functionality is managed through an easy-to-use Web interface. This interface is accessible from any Java-enabled Web browser without having to install any additional software. The Web interface is accessed through an SSL secured connection, making it safe to manage the PureDisk Storage Pool from a remote location.

An administrator defines the visible scope and the functional level for each user that logs on to the Web interface. The scope can be limited per client, group of clients, or location. The functional level can allow one user to only restore files while another user can be given the right to manage the configuration of a Storage Pool.

PureDisk agent-based restores are scheduled or initialized via the Web interface. Operators or users with the required access rights can

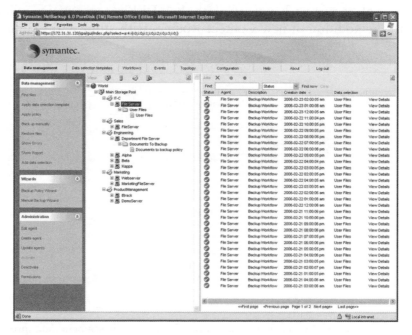

Figure 10. The NetBackup PureDisk Web-based interface.

manage agent-based restores from any location. Because PureDisk is disk-based, a restore in a remote location can be initiated centrally with no manual intervention onsite.

Web restore interface

In addition to agent-based backup-and-restore functionality, the Web-based search tool allows the user to find old files or file versions quickly and download these files via HTTP to any location without the need for a PureDisk agent.

Virtual file system interface

PureDisk also can be accessed through the PureDisk virtual file system interface. The PureDisk virtual file system interface is a CIFS interface that represents Metabase information in network shares. These network shares can be accessed from any CIFS-capable client.

The information presented in the CIFS mount point is filtered metadata. Filters can be defined based on client information, file type, path inclusions and exclusions, and source system presence time. For example, an operator can define mount points in the virtual file system that represent client data as it was present during the last backup, one week ago, and one month ago. If a file is copied from such a virtual mount point, the virtual file system will decompress, de-encrypt, and reassemble the file on the fly based on the data from the Metabase and the file segments from the Content Router(s).

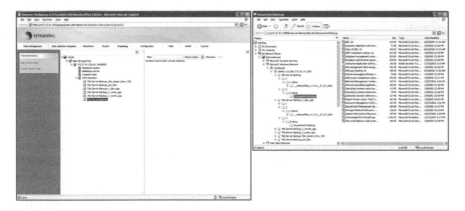

Figure 11.   The NetBackup PureDisk virtual file system interface.

Parallel processing

Backup processes are distributed to individual clients at both central and remote sites throughout the entire enterprise network. Scalability is guaranteed because large portions of the backup process for each client is performed locally on the system by an agent. This allows backups of hundreds of clients to run simultaneously.

Each client will only send the globally unique file segments to its Storage Pool. Even when hundreds of agents are performing backups, the volume of backup data on the network remains limited. Also, the Storage Pool needs to accept only a minimal volume of data. Thus, the parallel processing of hundreds of agents will not flood the network, nor will it stress the Storage Pool servers. The result is a highly scalable backup environment that is virtually insensitive to the number of clients and remote sites in your business IT environment.

Backup data encryption

The PureDisk agent optionally encrypts every file and file segment that it sends to the Storage Pool. Encryption of the data before sending it over the WAN to the Storage Pool ensures that the data is secure at all times—the data is unreadable and unusable by any hacker who may attempt to tap the data stream.

All encrypted file segments that are received by the Storage Pool are stored to the Content Router disks in their encrypted form. No description is performed, resulting in end-to-end data security. The disk architecture eliminates the risk of accidental loss of tapes, and unauthorized personnel cannot gain access to the data stored in the disk-based Storage Pool.

Policy-driven retention and data removal

By default, PureDisk will retain all versions of all protected and migrated data in the Storage Pool. Storage administrators can define policies that will remove obsolete data from the storage system. These policies can be applied on any selection of data sources. Removal policies can use time (older than), version (last n versions) and file attributes (extension, size) to filter the files to be removed.

Storage pool replication and disaster protection

Because of its unique content routing architecture, PureDisk can replicate its storage to multiple local and/or remote storage devices. Content optimization will keep the required bandwidth and the replication
time to a minimum. By replicating to Content Routers and storage
devices in different locations, an additional layer of protection is added
to safeguard against potential disaster at the main site. The complete
PureDisk architecture can be rebuilt and all the machines under backup
control can be restored from the data at the remote site.

Replication to the remote site can be performed in asynchronous
mode. This guarantees that those backups are performed in the LAN
for optimal backup performance and a short backup window, while
the offsite replication can take place at a slower rate.

CONCLUSION

- The fact is that more data will be generated and held on disparate
  computer systems. Organizations of all sizes need to ensure that their
  data is properly secured with the ability to store, access, recover, and
  ultimately remove it. Pressure from legislation, combined with press-
  ing business needs, require that data protection policies of proven
  benefit be in place.
- Many of the existing data-protection strategies are based on obsolete
  business practices such as a 10-hour working day and overnight
  backup windows. The modern world of 24-hour-a-day industry
  and electronic transactions has forced an increased reliance on data
  systems.
- NetBackup PureDisk combines breakthrough technology and modern data protection concepts to put the brakes on the exponential
  increase of backup data and bandwidth. It uses fingerprint technology to distinguish unique files and file segments from redundant ones
  and incorporates all systems in the data management process, even
  those on remote sites.
- Many organizations are in the process of deploying data protection
  solutions to solve shortterm problems. Many of these solutions
  revolve around one vendor or hardware platform. PureDisk offers
  a data protection infrastructure that provides a long-term solution
  with a predictable cost model—a solution with flexibility that
  embraces any hardware vendor or storage technology.

- You can reduce remote-office backup costs using the PureDisk policy-driven disk-to-disk backup technology that automates remote backup-and-restore tasks. Backup data is centralized using existing network connections, so manual media handling and shipments are no longer required.
- NetBackup PureDisk is a secure backup solution that only transmits and stores data after it has been encrypted. PureDisk components and the PureDisk interfaces use SSL-secured connections to communicate.
- The NetBackup PureDisk solution is able to adapt to emerging technology, legislation, and business processes. Content routing and fingerprint technology fundamentally changes the way data is handled and enables the kind of information lifecycle management that is needed now and in the foreseeable future.

ABOUT SYMANTEC

Symantec is the world leader in providing solutions to help individuals and enterprises assure the security, availability, and integrity of their information. Headquartered in Cupertino, Calif., Symantec has operations in more than 40 countries. More information is available at www.symantec.com.

For specific country offices and contact numbers, please visit our Web site. For product information in the U.S., call toll-free 1 (800) 745 6054.

Symantec Corporation
World Headquarters
20330 Stevens Creek Boulevard
Cupertino, CA 95014 USA
1 (408) 517 8000
1 (800) 721 3934
www.symantec.com

Index